Platelets and Aspirin-Induced Asthma

Platelets and Aspirin-Induced Asthma

Pathogenesis and Melatonin

Helen V. Evsyukova

Edited by

M. Joseph
Former Director of Research from Centre National de la Recherche
Scientifique, Institut Pasteur de Lille, France

AMSTERDAM • BOSTON • HEIDELBERG • LONDON
NEW YORK • OXFORD • PARIS • SAN DIEGO
SAN FRANCISCO • SINGAPORE • SYDNEY • TOKYO
Academic Press is an imprint of Elsevier

ELSEVIER

Academic Press is an imprint of Elsevier
The Boulevard, Langford Lane, Kidlington, Oxford, OX5 1GB, UK
225 Wyman Street, Waltham, MA 02451, USA

First published 2014

British Library Cataloguing-in-Publication Data
A catalogue record for this book is available from the British Library

Library of Congress Cataloging-in-Publication Data
A catalog record for this book is available from the Library of Congress

ISBN: 978-0-12-800033-5

This book has been manufactured using Print On Demand technology. Each copy is produced to
order and is limited to black ink. The online version of this book will show color figures where
appropriate.

CONTENTS

AA	arachidonic acid
ACTH	Adrenocorticotropin
ADP	adenosine diphosphate
AERD	aspirin-exacerbated respiratory disease
AIA	aspirin-induced asthma
AIANE	European Network on Aspirin-Induced Asthma
ALAT	alanine aminotransferase
AOS	antioxidant system
APUD	amine precursor uptake and decarboxylation, a specialized highly organized system
ASA	acetylsalicylic acid
ASAT	aspartate aminotransferase
ATA	aspirin-tolerant asthma
AVRI	acute viral respiratory tract infection
BAL	bronchoalveolar lavage
CIC	circulating immune complexes
CNS	central nervous system
COX	cyclooxygenase
cAMP	$3'$-$5'$-cyclic adenosine monophosphate
cGMP	$3'$-$5'$-cyclic guanosine monophosphate
CD4+	defines MHC class ll-restricted T-cell subsets
CD8+	defines MHC class I-restricted T-cell subset; present on NK cells
C3a	complement fragment 3a (anaphylatoxin)
DNA	deoxyribonucleic acid
DNIES	diffuse neuroimmunoendocrine system
EEG	electroencephalography
EX	eotaxin
$FEF_{50\%}$, $FEF_{75\%}$	forced expiratory flow at 50% and 75% of FVC
FEV_1	forced expiratory volume in 1 s
FGF	fibroblast growth factor
GC	glucocorticoid hormones
GM-CSF	granulocyte-macrophage colony-stimulating factor
G_i, G_o, G_s	G-proteins

HETEs	hydroxyeicosatetraenoic acids
HIOMT	hydroxyindole-O-methyltransferase
5-HT	serotonin
HPA	hypothalamic–pituitary–adrenal axis
IFN-γ	interferon-gamma
IgA, IgE, IgG	immunoglobulins A, E, G
IGF	insulin-like growth factor
IL	interleukin
LO	lipoxygenase
LPO	lipid peroxidation
LT	leukotriene
LX	lipoxin
MHC	major histocompatibility complex
MT	melatonin
α-MT6s	6-sulfatoxymelatonin
NADPH	reduced nicotinamide adenine dinucleotide phosphate
NANC	nonadrenergic, noncholinergic nervous system
NARES	nonallergic rhinitis with eosinophilia syndrome
NAT	N-acetyltransferase
NK	natural killer
NO	nitric oxide
NOS	nitric oxide synthase
cNOS	constitutive form of NOS
iNOS	inducible form of NOS
eNOS	endothelial NOS
nNOS	neuronal NOS
NSAIDs	nonsteroidal anti-inflammatory drugs
ONOO$^-$	peroxynitrite
PAF	platelet-activating factor
PDGF	platelet-derived growth factor
PEF	peak expiratory flow
PG	prostaglandin
PRP	platelet-rich plasma
RANTES	regulated on activation, normal T-cell expressed and secreted (a member of the IL-8 supergene family)
RNA	ribonucleic acid
RV	residual volume
SCN	suprachiasmatic nuclei
sG_{aw}	specific airway conductance

SNAS	5-sulfatoxy-*N*-acetyl-serotonin
SRS-A	slow-reacting substance of anaphylaxis
T_{h1}, T_{h2}	subsets of helper T-cells
TLC	total lung capacity
TNF	tumor necrosis factor
Tx	thromboxane
VC	vital capacity

H.V. Evsyukova is a professor of Medicine at the Department of Hospital Therapy, Medical Faculty of Saint Petersburg's State University (Saint Petersburg, Russian Federation). She was qualified in medicine from the Saint Petersburg State Medical University – a university named after academic I.P. Pavlov. She was trained in internal and respiratory medicine and undertook postdoctoral training at the departments of Hospital Therapy and Pathological Physiology of the I.P. Pavlov State Medical University. She was awarded PhD and DSc degrees in specialities of pulmonology and pathophysiology in 1992 and 2002. She is a member of the European Respiratory Society (ERS) and European Network on Aspirin-Induced Asthma (AIANE). She has authored and coauthored more than 100 scientific articles, reviews, and book chapters including the monograph "Differential diagnostic of lung diseases (selected chapters)," chapters in books "Allergology," "Inflammation mechanisms of bronchial tree and lungs, and anti-inflammatory therapy" (editor G.B. Fedoseev), "Bronchial asthma" (editors G.B. Fedoseev and V.I. Trofimov), "Hemostasis system" (editor N.N. Petrishchev), and also the chapters "Neuroimmunoendocrinology of respiratory organs" in the "Manual on neuroimmunoendocrinology" (editors M.A. Paltsev and I.M. Kvetnoy) and "Aspirin-intolerant asthma" in the book "Many faces of bronchial asthma: understanding, diagnostics, treatment and preventive care" (editors G.B. Fedoseyev, H. Upur, V.I. Trofimov, M.A. Petrova). She is the author of invention patent "Diagnostic of aspirin-induced asthma." The scientific contribution was noted with including her biography into the special edition "Who's who in medicine and healthcare 2009–2010."

One of the most amazing observations brought out by the research in cellular immunity during the last three decades was the promotion of blood platelets to the status of effectors in the destruction of several pathogens and their involvement in the mechanisms of various inflammatory disorders. It is now widely accepted that, besides their classical role in thrombosis and hemostasis, blood platelets participate in other physiological processes, such as the initiation of tissue repair after injury, or the defense against bacteria, tumors, and parasites. These properties have led to the concept that platelets themselves can be considered as inflammatory cells [247]. Initially, it was suggested that the thrombocyte involvement in inflammatory reactions was a consequence of their passive role of targets for leukocyte mediators released in various circumstances of cell activation. Among these mediators, platelet-activating factor, interferon-γ, tumor necrosis factors, interleukin 6, C-reactive protein, or substance P were proposed as agents with stimulating properties for platelets. However, it appeared that these blood elements had an active role on their own by excreting specific factors *in vitro* and *in vivo*, such as platelet factor 4, β-thromboglobulin, platelet-derived growth factor, or histamine release-inducing factors, all with potent amplifying functions for basophils, mast cells, and other inflammatory cells. Finally, in the context of allergies, it was shown that platelets could also be seen as directly involved in the immunological mechanisms of immediate hypersensitivity reactions, and thus in the inflammation associated with allergy, partly as a consequence of the presence of receptors on their surface for IgE (FcϵRI [248] and CD23 [246]).

In fact and unexpectedly, the demonstration of the binding of IgE to platelets led to the observation of a participation of platelets to aspirin-induced asthma (AIA). When analyzing cytotoxic functions of various cells against the parasite *Schistosoma mansoni*, we observed that the larvae of this pathogen were efficiently killed *in vitro* by blood platelets, either passively sensitized by IgE-specific antibodies when isolated from healthy donors or purified from the blood of patients infected with the parasite. Catalase and superoxide dismutase were good inhibitors of the platelet cytocidal activity against the parasite, focusing on a potential generation of oxygen-dependent free radicals among the mediators produced by platelets in these effects. When using platelets from healthy donors or from aspirin-tolerant asthmatics, aspirin and nonsteroid anti-inflammatory drugs (NSAIDs)—inhibitors of cyclo-oxygenase—inhibited the IgE-mediated cytocidal stimulation of platelets. With platelets from aspirin-intolerant patients, not only aspirin and NSAIDs were unable to inhibit the IgE-dependent cytotoxicity but they also increased the cytocidal effect. It appeared that these drugs induced directly the generation of cytotoxic mediators by AIA platelets [12]. Conversely, no other leukocyte population from aspirin-intolerant asthmatics could kill parasite larvae *in vitro* in the presence of acetylsalicylic acid or NSAID.

This main focus on platelets, if not exclusive, identified partners in some physiological processes, outside their classical role in hemostasis, which deserves interest. These blood elements also appeared as the crucial actors in another effector mechanism, namely in the activity of an antifilarial drug, diethylcarbamazine (DEC). Familiar for more than half a century, DEC presents a marked contrast between an extremely rapid action *in vivo* and the absence of any significant activity on microfilaria *in vitro*. In 1987, we demonstrated that the effect of DEC is mediated by blood platelets with the additional triggering of a filarial secretory antigen [82]. The killing mechanism is antibody independent and involves the participation of free radicals. Here again, only platelets were active *in vitro*, when purified from the blood of healthy donors 3 h after the ingestion of DEC. Fifteen days after oral administration of the drug, platelets were still active against microfilaria, whereas free DEC was no more detectable in the serum after 48 h. This suggested that the target of DEC *in vivo* was platelets and, probably, their bone marrow precursors, since the effect of DEC outlasted the average platelet lifetime of 9–10 days.

The experiments summarized above illustrate the ability of platelets to mediate profound and efficient processes in the human physiology. In this perspective, it is the merit of Helen Evsyukova to have opened a new chapter in the involvement of blood platelets in the pathophysiology of AIA, especially through the regulating role of melatonin. The researches she has initiated with her collaborators for decades have opened interesting insights into the role of melatonin as a regulator of biological rhythms, as endogenous antioxidant, and in modulating nitric oxide production and arachidonic acid metabolism, a biochemistry which has particular consequences in association with blood platelets. The present book offers an exhaustive overview of AIA, together with a precise presentation of the molecular and cellular dysfunctions sustaining that disorder. At a time when acetylsalicylic acid, far from being out of date, becomes a prerequisite in the treatment of developing pathologies in our societies, especially in cardiac and vessel medicine, the researches reported here extensively open a large panel of trails to be followed by scientists in AIA. I thank Helen for her confidence in asking me to write the foreword of this significant opus.

Now retired from the scientific activity, my participation in the book of Professor Helen Evsyukova is most probably my ultimate contribution to the medical research. I thank Helen Evsyukova to have allowed me to take this opportunity to express here my profound gratitude to a man who was my master in research, in sustaining my career for years, and by associating me to several of his intuitions and missions, esteemed Professor André Capron. He passed on to me his enthusiasm for the challenges of the research and gave me his constant and almost paternal friendship. I am forever grateful to him. Among many pioneering discoveries, it is in his laboratory that the interaction of aspirin with blood platelets in aspirin-intolerant asthmatics was described for the first time. Finally, I would share this acknowledgment with his wife, dear Professor Monique Capron. Thank you to both of you for everything.

Michel Joseph

Former Director of Research from
Centre National de la Recherche Scientifique
Institut Pasteur de Lille, France

INTRODUCTION

Salicylates have been used in clinical practice for over 2400 years. It was Hippocrates (around 400 BC) who prescribed an infusion made of the white willow bark to relieve pain and fever. In 1829 Leroux discovered willow's active ingredient, salicin; later on Kolbe isolated and extracted salicylic acid, and in 1897 Felix Hoffman was able to synthesize acetylsalicylic acid (ASA), and the Bayer Company registered Aspirin as a trademark [270]. By the start of the twentieth century, aspirin had become the most popular drug throughout the world.

Shortly after its being introduced into clinical practice, aspirin was found to cause an acute bronchospasm. The association of aspirin intolerance, bronchial asthma, and nasal polyposis was first described by Widal in 1922. In 1968 Samter and Beers [496] named this illness "asthmatic triad." For over five decades since the first description of this symptom complex, it has inspired an ever-increasing interest among researchers, as the clinical course of this disease may be severe and lead to disability of patients. It creates serious medical and social problems, that can only be resolved through further insight into its pathogenesis.

In the 1970s, John Vane [603] discovered a mechanism of anti-inflammatory action of acetylsalicylic acid and opened a new way to understanding the mechanisms of aspirin hypersensitivity. Polish researchers Szczeklik and Gryglewski [557] have demonstrated that it is not only aspirin, but other nonsteroidal anti-inflammatory drugs (NSAIDs) inhibiting cyclooxygenase may cause adverse reactions in respiratory tracts. In 1993 a European Network on Aspirin-Induced Asthma (AIANE) [568] was set up to support studies in epidemiology, mechanisms of NSAIDs intolerance symptoms development in bronchial asthmatics, and methods of the disease diagnosis and treatment.

Epidemiological studies in different countries show a high prevalence of aspirin-induced asthma (AIA). However, the occurrence of case finding depends on methods of aspirin intolerance diagnosis and on cohorts examined. Thus, Szczeklik and Nizankowska [561] have pointed out that in European countries aspirin intolerance among adults with bronchial asthma, according to anamnesis, occurs in 3–5% of cases, while in the event of oral aspirin provocation test this number goes up to 19%, and in bronchial asthma patients with associated polyposis rhinosinusitis aspirin intolerance is revealed in 34% [515]. Similar data on the prevalence of AIA can be found in studies by different authors from Australia [243], Venezuela [502], Italy [509], and other countries [370].

The frequent association of asthmatic syndrome and aspirin intolerance with the pathology of upper airways (nasal polyposis, vasomotor rhinitis, chronic hyperplasic eosinophilic sinusitis) was deemed to be a reason for some researchers to view AIA as a part of a single complex of symptoms which was called aspirin-exacerbated respiratory disease (AERD) [86,546,627].

Nevertheless, taking into account the fact that upper airway pathology is absent in one-third of patients with bronchial asthma and NSAIDs intolerance, this respiratory disease is also called aspirin-sensitive asthma or AIA. According to Global Strategy for Asthma Management and Prevention [199], AIA is determined as a specific occurrence of bronchial asthma which requires additional methods of diagnosis and treatment. Russian pulmonologists view AIA as a variant of clinical picture and pathogenesis of bronchial asthma [94,179].

The results of our studies and extensive clinical experience of dealing with such patients at the Hospital Therapy Department of St. Petersburg State Medical University named after I.P. Pavlov (St. Petersburg, Russia) allow us to view AIA as a systemic pathology different from bronchial asthma without aspirin intolerance, to apply new approaches to evaluation of its clinical presentation and pathogenesis, and to propose new methods of diagnosis and treatment.

CHAPTER 1

Clinical Picture of Aspirin-Induced Asthma (AIA)

1.1 CLINICAL RISK FACTORS AND COURSE OF ASPIRIN-INDUCED ASTHMA

Our findings have demonstrated that aspirin-induced asthma (AIA) is a disease with a clear and definite symptomatology, which differentiates it from aspirin-tolerant asthma (ATA). A particular feature of AIA is that it occurs in age periods when the human organism undergoes a hormonal alteration—in prepubertal children, women in their thirties, and men in their fifties or sixties. Women amount to 75% of cases.

Prior to the first shortness of breath, most patients suffer pathology of the upper respiratory tract (see Table 1.1).

The same peculiarities have been noted by other researchers as well [434,567,572,595]. We have also demonstrated that along with the occurrence of upper airway pathology, the initial symptoms of lung disease are observed that may point to changes in the reactivity of the bronchial tree: appearance of dry paroxysmal cough with blennoptysis and expiratory dyspnea in response to irritant inhalation, cold air, as well as in case of changing weather conditions, or exercise. Patients are often diagnosed with chronic asthmatic bronchitis. At the same time, some of them have noted the occurrence of angioedema in response to unknown exogenic factors. It is often a case that shortly before the illness, the patients may have suffered concussion of the brain, had a vaccination, or been under a significant amount of stress.

The first asthma attack may be provoked by contact with infectious or noninfectious allergens, operative intervention in accessory sinuses of the nose, inhalation of cold air or irritant odors, or exercise. The illness in women often starts to develop in the climacteric period.

As a rule, asthma attacks in response to aspirin or other nonsteroidal anti-inflammatory drugs (NSAIDs) occur in a patient who already has asthma and/or nasal polyps with chronic rhinosinusitis. The intake of these medications, however, initiates the asthmatic syndrome

Table 1.1 Prevalence (%) of Different Diseases Prior to the Appearance of the First Attack of Breathlessness in Patients with AIA

Disease	Groups		
	AIA	ATA	P
Urticaria	23.9 ± 3.9	27.5 ± 4.4	>0.05
Neurodermitis	7.7 ± 2.9	4.9 ± 3.2	>0.05
Angioedema	17.9 ± 3.5	6.9 ± 3.2	<0.05
Vasomotor rhinitis	70.9 ± 4.2	44.9 ± 4.8	<0.001
Nasal polyposis	30.2 ± 4.3	2.9 ± 3.1	<0.001
Chronic asthmatic bronchitis	69.7 ± 4.2	52.2 ± 4.7	<0.01
Chronic mucopurulent bronchitis	16.8 ± 3.4	24.0 ± 4.2	>0.05

development in every one of three patients. Within an hour of aspirin ingestion, the patient may suffer an asthma attack, followed by rhinor-rhea, conjunctival injection, and flushing of the upper body, mostly the head and neck. In some patients, the asthma attack is followed by hypotonia, hypersalivation, nausea, and vomiting, as well as pains in the upper abdomen. All these reactions are extremely dangerous for the patient's life, as they may rapidly lead to shock, loss of conscious-ness, or even death.

After the occurrence of NSAIDs intolerance, the course of asthma changes: the seasonality and acuity of asthma attacks disappears, there is a sense of permanent congestion in the chest, and the efficiency of ordinary bronchodilators decreases. Severe exacerbations of the disease are observed more than four times a year, irrespective of a season, and most patients with AIA (unlike aspirin-tolerant patients) react to a wider range of external factors (Table 1.2).

It should be noted that the asthmatic syndrome in AIA patients develops against the background of the already-existing imbalance of the organism's functional systems. Thus, in most patients, along with the signs of typical atopy, there can be observed a link between the asthma attacks and acute viral respiratory tract infection (AVRI), pneumonia, or chronic infection activation (sinusitis, cholecystitis, pancreatitis, etc.), which may indicate a disorder of the immune sys-tem of such patients. The latter by anamnesis may also show various benign tumors. Thus, 71% of AIA patients have nasal polyps, 18.8% of women have hysteromyoma, and 36% of women have indications for myomectomy or hysterectomy. The predisposition of patients

Table 1.2 Prevalence (%) of Factors Provoking Recurrent Asthma Attacks in Patients with AIA

Factor	Groups		
	AIA	ATA	P
Contact with allergens (food, dust, pollen, epidermal sensibilization)	88.4 ± 2.9	86.1 ± 3.0	>0.05
Acute viral respiratory tract infection	79.7 ± 3.7	75.2 ± 4.0	>0.05
Acute bacterial disease or exacerbation of chronic inflammatory disease	62.8 ± 4.5	46.5 ± 5.0	<0.05
Cold air or irritant inhalation	93.3 ± 2.9	65.8 ± 4.5	<0.001
Physical exercise	73.9 ± 4.0	57.8 ± 4.7	<0.05
Changing weather conditions	55.1 ± 4.6	33.7 ± 4.7	<0.01
Psychological stress	75.8 ± 3.9	51.4 ± 4.9	<0.001
Premenstrual phase of menstrual cycle	22.0 ± 3.8	6.9 ± 3.2	<0.01
Taking NSAIDs	86.7 ± 3.1	0.0 ± 3.2	<0.001
Taking other medications	59.7 ± 4.5	40.4 ± 4.7	<0.01
Eating products containing natural salicylates	46.1 ± 4.6	5.0 ± 3.2	<0.001
Drinking strong alcoholic beverages	22.3 ± 3.9	14.9 ± 3.5	>0.05
Mycetogenic sensibilization	46.6 ± 4.6	31.1 ± 4.5	<0.05

with AIA to multiple polyposes (in the urogenital and gastrointestinal systems) has been noted by other authors [448]. The disorder of immune system functions in AIA has been also evidenced by our data on reduced resistance to AVRIs, existence of chronic infectious inflammatory processes, mycetogenic sensibilization, as well as the onset of asthmatic syndromes after immunization. Furthermore, we have found out that every second patient with AIA has a chronic persistent chlamydial infection caused by *Chlamydia pneumoniae*. This is confirmed by the fact that 38% of patients with AIA have IgG titers $\geq$1:32 to *C. pneumoniae* along with an increase of circulating immune complexes (CIC) in blood (>0.1 units) and a decrease of the monocytes migration index (<30%).

It is known that the development of viruses and intracellular micro-organisms' persistence, as well as the disorder of the organism's antitumor defense, is based on the immune system deficiency, including the functional deficiency of T cells and natural killer (NK) cells [430,634]. Some authors point to a decrease of absolute and relative number of T lymphocytes and especially T helpers in patients with AIA and a change of their functional state, manifested in the distortion of T-lymphocytes response to serotonin and normal reaction to histamine, adrenalin, and theophylline [552]. It is believed that subjects with AIA have a deficit of NK cells as well as an imbalance of cellular subpopulations sensitive to monoamines [551]. Thus, the existence of immune system deficiency in AIA has been confirmed by other researchers.

Most patients with AIA at the beginning of the disease demonstrate symptoms which indicate to dysfunctions of central nervous system (CNS) early. Most patients with AIA, much more frequently than those with ATA, note the connection of asthma attacks with emotional suffering or psychological stress (Table 1.2). A dependency of the asthma attacks from psychoemotional suffering is diagnosed in 78.6% of young patients, and its frequency remains high in all ages. Anxiodepressive and asthenoneurotic syndromes are more frequently observed in patients with AIA, which is an evidence of increased anxiety, immature value system, and a low capability of independent decision making. In view of the evolutionary theory, such an organism possesses the least perfect mechanisms of environment adaptation [280]. Indeed, according to our data, in 18.3% of patients with AIA, the onset of the disease is linked with a psychoemotional trauma. This data which correlates with recently published materials of the European Network on Aspirin-Induced Asthma (AIANE), that demonstrate that psychological stress has been a triggering factor of the onset and exacerbations of the disease in 240 of 365 AIA patients [386]. Furthermore, the occurrence of nonallergic rhinitis with eosinophilia syndrome (NARES), which is viewed by Moneret-Vautrin et al. [366] as a preexisting disease of aspirin triad, is also linked to the existence of stress situations in the anamnesis (mourning, labor conflicts, divorce, unemployment, etc.) in half of the patients with AIA.

Among the factors that provoke recurrent asthma attacks, most patients with AIA point to cold air, irritant inhalation, physical exercise, and change of weather conditions (Table 1.2). It all of which are evidence of an airway hyperresponsiveness. It is known that in the genesis of airway hypersensitivity and hyperresponsiveness, which are the characteristic functional abnormality of bronchial asthma, a significant role belongs to the imbalance of exciting and inhibitory impacts of different elements of the vegetal nervous system. It is a common view that in patients with bronchial asthma there is an increased effect of acetylcholine parasympathetic neurotransmitter and dysfunction of adrenergic system—that is, an increase of α-receptors sensitivity and a decrease of β-receptor sensitivity [376,490,652]. As noted earlier, a change in the reactivity of airways is observed in patients prior to the appearance of typical asthma attacks, which is an evidence of early dysfunctions of vegetal regulation in AIA. This is also confirmed by the data of Moneret-Vautrin et al. [366] who have found a dysfunction of the autonomic nervous system in patients with NARES syndrome. The authors have revealed an increased sensitivity to adrenergic agents at the level of β_2-receptors of large vessels and α_1-receptors of small skin vessels in case of intradermal injection of papaverine hydrochloride causing edema and vasodilatation in healthy persons. At the same time, the activity of lymphocyte β-adrenoreceptors was not significantly different from those in healthy subjects [630]. This fact confirms that the dysfunction of central regulatory mechanisms in such patients play a certain role in the formation of an initial dysfunction of the autonomic nervous system. Thus, in patients with the asthmatic triad there were some changes recorded by electroencephalography (EEG) that pointed to a dysfunction of intracentral connections according to "diencephalic variant" and their relation to a change in the airway sensitivity threshold and responsiveness [316].

Along with dysfunctions in the nervous and immune systems of AIA, we have revealed early occurrences of functional disorders in the endocrine system. Thus, in 50% of female patients of young age, the exacerbation of asthma occurs in the premenstrual period. The studies by Markov [336], Kagarlitskaya [250], and Sabry [491] have also demonstrated a high percentage of patients with AIA among women suffering from premenstrual asthma attacks. Our analysis of the frequency of different past diseases has shown that 26% of female patients with AIA had menstrual disorders, miscarriages, and habitual noncarrying of pregnancy. Moreover, we have noted an early onset of menopause in these patients, at 46 years on average, as compared to 50 years in patients with ATA. Additionally, 36% of women with AIA had clinical symptoms of menopause after hysterectomy due to uterine fibroids. Besides of the high frequency of reproductive function disorders, we have found that patients with AIA have thyroid pathology twice as often as patients with ATA. Functional deficiency of many organs in the endocrine system is evidenced by the early development of asthma attacks after the glucocorticoid (GC) hormones' cessation or the decrease of its dose. According to our data, 28.6% of patients aged

16–29 years with AIA had a steroid dependency of the asthma course, compared to 2.7% of ATA patients. Its frequency remains almost at the same level (19%) in middle and mature years, and decreases to 12.4% among elderly patients with AIA. The coinciding ages (16–29 years) of GC's dependency and reproductive function disorders in patients with AIA may confirm that the formation of GC dependency is caused by damage to central elements that regulate the GC's function of the adrenal cortex.

Patients with AIA often have chronic comorbidity of other organs and systems, mainly diseases of upper respiratory tracts. This has been noted by other researchers who pointed to high frequency (60%) of nasal polyposis and chronic eosinophilic rhinosinusitis [20,60,139,434,436,562], which correlates with our data (71%). AIA patients have also patients frequently also have inflammatory diseases of nasal sinuses (chronic maxillitis, sinusitis, ethmoiditis).

In addition to the pathology of upper airways, in most patients with AIA we have noted comorbidity of the gastrointestinal tract (chronic gastritis, peptic ulcer, cholecystitis, biliary dyskinesia, Gilbert's disease). Additionally, in 26.4% of patients we have observed varicosity of lower limbs, which indicates the involvement of the vascular system in the pathologic process.

The analysis of genetic predisposition to various diseases has not revealed any specific features in the patients with AIA and ATA. At the same time, close relatives of AIA patients suffered from tuberculosis of respiratory organs three times more often and showed various reactions to aspirin intake nine times more often. Studies conducted in Europe demonstrated that 5.2% of patients with AIA have indications to NSAIDs intolerance in the family anamnesis [135].

AIA is characterized by a severe disease course irrespective of NSAIDs intake, requiring a constant use of peroral corticosteroids [135,174,343,509,569]. According to our data, it is observed 2.5 times more often than in patients with ATA.

Thus, the analysis of clinical manifestation of bronchial asthma in patients with acetylsalicylic acid (ASA) and NSAIDs intolerance, and in patients without such intolerance, has shown that in AIA patients, the disease development from the earliest stage is characterized by the involvement of the organism's nervous, endocrine, and immune systems in the pathologic process, which determines the severity of the patients' state. An asthma attack that occurs with that background may be provoked by an overstrain of the CNS (significant emotional stress or craniocerebral trauma), immune system (vaccination or AVRI), or endocrine system (menopause, etc.), as well as by aspirin intake, and leads to the development of an asthmatic reaction and bronchial asthma diagnosis. Our studies, however, have shown that asthmatic reaction in this group of patients is only one of numerous clinical manifestations of the disease which forms the "aspirin-sensitive asthma syndrome." The rapid severity progression of respiratory disease is a consequence of a huge damage not only to the upper and lower airways but to the vascular system of the lungs and other organs as well.

1.2 PECULIARITIES OF PULMONARY FUNCTION, LABORATORY DATA, AND PULMONARY CIRCULATION STATE IN AIA PATIENTS

Comprehensive examination upon admission to hospital, including chest and nasal sinuses radiography, lung function testing, clinical and biochemical blood tests and sputum cytology, has allowed determination of the peculiarities of laboratory and functional investigation results which are specific to AIA patients. X-ray studies have shown that most patients with AIA have a total shadowing of nasal sinuses as a result of mucosal thickening due to its edema.

A comprehensive examination of the lung function before and after the administration of a bronchodilator (e.g., 200 µg inhaled salbutamol) also revealed significant differences between the bronchial asthma patients with and without aspirin intolerance. Thus, all the patients with bronchial asthma proved to have airway obstruction, and the intensity of airflow limitation was different at the level of proximal and distal bronchi (Table 1.3). It is clear from Table 1.3 that a major part of the revealed defects in AIA patients is related to airflow limitation at the level of distal bronchi both before and after the inhalation of β_2-adrenergic agonist (Table 1.4). Most of the AIA patients (71.4%), in contrast to 57.3% of the ATA patients ($P < 0.05$), demonstrated an evident degree of bronchospasm.

The obtained data demonstrate the existence of bronchial obturation syndrome in AIA patients, which is mostly manifested in the distal end of the bronchial tree, and defects of microcirculation in the lungs.

Table 1.3 Pulmonary Function Indices Before Inhalation of β_2-Adrenomimetic in AIA Patients

Indices	Group		
	AIA	ATA	P
VC (% pred)	98.0 ± 2.0	93.0 ± 2.0	>0.05
RV (% pred)	184.0 ± 6.0	172.0 ± 6.0	>0.05
TLC (% pred)	125.0 ± 2.0	125.0 ± 9.0	>0.05
FEV$_1$ (% pred)	73.0 ± 2.0	75.0 ± 2.0	>0.05
sG$_{aw}$ (cm H$_2$O^{-1} s^{-1})	0.083 ± 0.003	0.093 ± 0.003	<0.05
PEF (% pred)	68.0 ± 2.0	75.0 ± 2.0	<0.05
FEF$_{50\%}$ (% pred)	35.0 ± 2.0	47.0 ± 3.0	<0.001
FEF$_{75\%}$ (% pred)	31.0 ± 1.0	44.0 ± 2.0	<0.001

Pulmonary function indices: VC - Vital capacity, RV - Residual Volume, TLC - Total Lung Capacity, FEV$_1$ - Forced expiratory volume in one second, sG$_{aw}$ - Specific airway conductance, PEF - Peak expiratory flow, FEF$_{50\%}$, FEF$_{75\%}$ - Forced expiratory flow at 50% and 75% of FVC

Table 1.4 Pulmonary Function Indices After Inhalation of β_2-Adrenomimetic in AIA Patients

Indices	Group		
	AIA	ATA	P
VC (% pred)	110.0 ± 2.0	103.0 ± 2.0	<0.01
RV (% pred)	143.0 ± 4.0	133.0 ± 4.0	=0.001
TLC (% pred)	120.0 ± 1.0	114.0 ± 1.0	<0.001
FEV$_1$ (% pred)	94.0 ± 2.0	93.0 ± 2.0	>0.05
sG$_{aw}$ (cm H$_2$O^{-1} s^{-1})	0.119 ± 0.004	0.136 ± 0.006	<0.05
PEF (% pred)	88.0 ± 2.0	92.0 ± 2.0	>0.05
FEF$_{50\%}$ (% pred)	55.0 ± 2.0	66.0 ± 3.0	<0.01
FEF$_{75\%}$ (% pred)	47.0 ± 2.0	61.0 ± 3.0	<0.001

The results of ventilation/perfusion lung scintigraphy, conducted earlier, have revealed capillary circulation disturbances. In the group of AIA patients, the totally preserved lung perfusion amounted to 27.8% ± 3.9%, while in ASA-tolerant patients with bronchial asthma it was 51.0% ± 8.0% ($P < 0.01$). In mostly affected areas, the capillary circulation in AIA patients remained at the level of 35.6% ± 7.6%, while in ATA patients it was 63.6% ± 11.1% ($P < 0.05$), and 55% of AIA patients had large areas with undetectable microcirculation [147,155,175,177]. The comparison of microcirculation and pulmonary function for each patient shows a high correlation between the reduction in pulmonary perfusion and the degree of airways obstruction. There is no such correlation in ATA patients.

As for the results of cytological examination of sputum and bronchial lavage, there is no significant difference between the AIA patients and the aspirin-tolerant patients with bronchial asthma.

The clinical blood analysis shows a specific state of hemopoiesis related to the generation of erythrocytes and megakaryocytes because AIA patients have shown an increased number of platelets $((255 \pm 7) \times 10^9$ cells $L^{-1})$ and decreased erythrocytes number $((3.5 \pm 0.1) \times 10^{12}$ cells $L^{-1})$ compared to ATA patients $((231 \pm 8) \times 10^9$ cells L^{-1} and $(4.6 \pm 0.1) \times 10^{12}$ cells L^{-1}; $P < 0.05$, respectively).

The data of laboratory and instrumental examinations of bronchial asthmatics show that AIA patients have microcirculation dysfunction connected with the pathology of platelet-vessel hemostasis. It is clinically manifested in mucosal edema of the nasal sinuses and bronchial tree, the development of vasomotor rhinitis, and bronchial obturation syndrome.

Thus, our studies demonstrate that AIA is a disease with a clinical symptomatology that is different from bronchial asthma without aspirin and other NSAIDs intolerance. It is characteristic for an AIA patient to combine pathologies of all functional systems of the body (nervous, immune, endocrine, hepatobiliary, hemopoiesis, hemostasis system, etc.), expressed in their functional deficiency. At the same time, ATA patients mostly demonstrate a combination of clinical and laboratorial indications, which points to hyperfunction of various organs and systems (atopic sensibilization with IgE hyperproduction, high erythrocytes and leukocytes in blood, increased activity of alanine aminotransferase (ALAT) and aspartate aminotransferase (ASAT), essential hypertensia, increased diastolic blood pressure, etc.). Therefore, AIA is a specific disease characterized by pathology and imbalance in all functional systems. Such dysfunctions can occur due to damage to the amine precursor uptake and decarboxylation (APUD) system (a specialized, highly organized cell system), whose main feature is the ability to uptake biogenic amines and decarboxylize them with subsequent production of biogenic amines and peptide hormones. The latter controls metabolic, secretory, immune, and other processes in the organism, providing the optimal level of a living system's homeostasis. Numerous recent studies have broadened our understanding of the structure and functions of the APUD system which, at present, is viewed as a diffuse neuroimmunoendocrine system (DNIES) [405].

CHAPTER 2

Diffuse Neuroimmunoendocrine System (DNIES) and its Role in the Human Body

2.1 DEFINITION

In the 1960s, British scientist A.G.E. Pearse [419] was the first to suggest that in the human organism there is a specialized, highly organized cell system with a primary function of producing peptide hormones and biogenic amines [288]. These cells have been called APUD cells, the name being derived from the acronym referring to "amine precursor uptake and decarboxylation." According to A.G.E. Pearse, these cells secrete biologically active substances that play the role of intercellular signaling and ensure the interrelated functioning of many organs and systems within the human body. Later on, numerous studies have demonstrated that in visceral organs there are nervous, endocrine, and immune cells that, together with APUD cells, produce numerous peptides and biogenic amines identical to those in the brain and in the central organs of the immune and endocrine systems. The same function is performed by nonendocrine cells: mast cells, NK cells, eosinophilic leukocytes, endothelial cells, monocytes, platelets, thymic epithelial cells, chondrocytes, osteocytes, placental trophoblasts and amnion cells, ovarian Leydig cells, endometrial cells, retinal photoreceptors and amacrine cells, Merkel cells in skin, Paneth cells, and macrophages.

The range of hormonal substances that are produced by these cells is extremely wide and includes serotonin, melatonin (MT), catecholamines, histamines, endorphins, endothelins, matrilysin, natriuretic peptide, vasoactive intestinal peptide, vasopressin, oxytocin, thymosins, insulin and insulin-like substances, somatostatin, prolactin, adrenocorticotropin (ACTH), leptin, and other molecules. Additionally, endothelial cells of vessels and platelets produce nitric oxide (NO), a gaseous hormonal substance that plays an important role in ensuring intercellular interactions [406].

Thanks to established facts, researchers have presently formulated a concept assuming the existence of DNIES in the organism, which integrates the nervous, endocrine, immune, and APUD systems through a common chemical mechanism of their functioning. Cells of this system produce signaling molecules providing a quick and adequate regulation of homeostasis in the live organism based on a "tuning-fork" mechanism, where an action of one substance leads to "a tuning-fork response and cascade expansion of perturbations" akin to "peptide–peptide" and "peptide–monoamine" interactions. This universal chemical language unites the systems controlling the organism's vital activity into a single functional mechanism regulating its functions [405].

MT is thought to be a hormone that unifies the DNIES, as it plays an important role in regulating the biological rhythms, coordinating the intercellular relationships and integrating the activity of all functional systems of the organism [289,301,408]. From a phylogeny point of view, MT is the most ancient hormone; its products have been discovered in many plants [470], confirming its important role in proving the vital activity of live systems.

Studying the role of MT in the pathogenesis of AIA is of a particular interest, as it has been found that one of its metabolites is a substance (N-acetyl-5-methoxy-kynurenamine) that is chemically similar to, and has the same effects as, ASA [256]. At the same time it has been demonstrated that MT is capable of regulating the expression of the 5-lipoxygenase (5-LO) gene and therefore controls leukotriene (LT) production in the organism [80].

2.2 MELATONIN (MT) AS DNIES HORMONE AND ITS ROLE IN REGULATING THE ACTIVITY OF THE ORGANISM'S FUNCTIONAL SYSTEMS

2.2.1 MT Synthesis and Metabolism in the Organism

It is known that MT is produced by the pineal gland located between the cerebrum and the cerebellum in the sulcus between the *colliculi superiores*. The main secretory elements of the pineal gland are pinealocytes [29,30,85].

A source of MT synthesis is tryptophan which comes to pinealocytes from the circulation and first turns into 5-hydroxytryptophan and then into 5-hydroxytryptamine (serotonin), from which MT is produced (Figure 2.1). N-acetyltransferase (NAT) and hydroxyindole-O-methyltransferase (HIOMT) are thought to be the most important enzymes involved in serotonin (5-HT) metabolism and MT synthesis [251,351,473]. NAT is the main enzyme that restricts the speed of MT

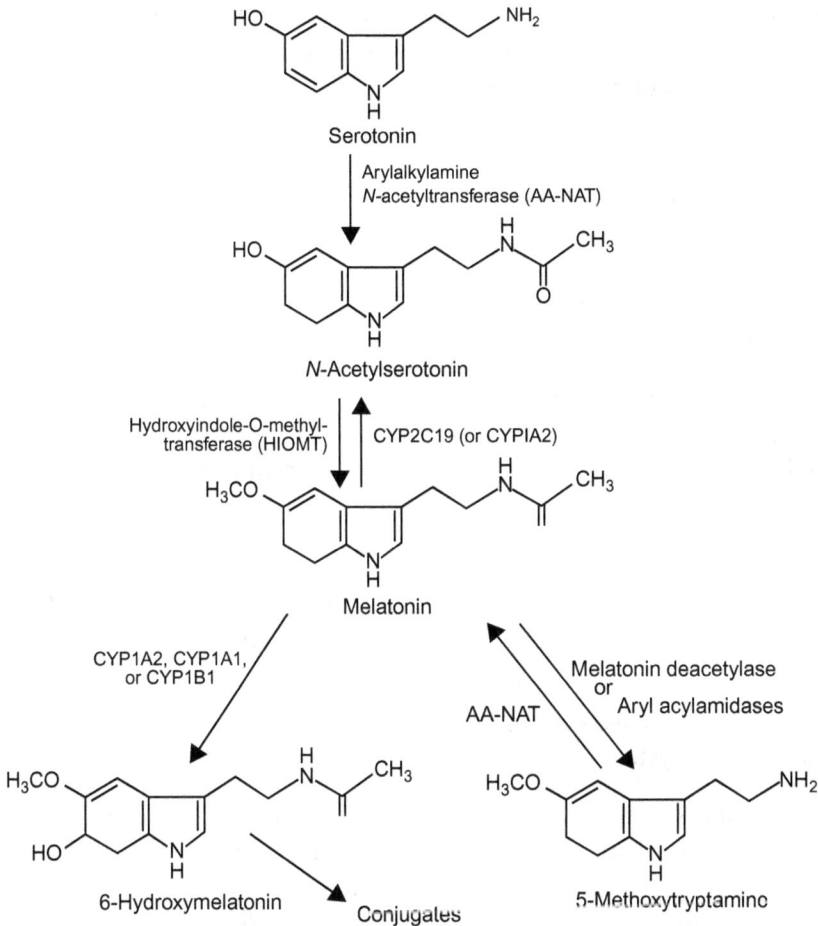

Figure 2.1 Formation of MT, its major pathways of indolic catabolism, and interconversions between bioactive indoleamines. CYP, cytochrome P_{450} isoforms (hydroxylases and demethylases). From Pandi-Perumal et al. [407]

formation. A gene has been cloned recently which regulates the NAT activity, located on the human 17q25 chromosome [103].

The MT production in the pineal gland is well studied. A nervous impulse from photoreceptor cells in the retina goes along the retino-hypothalamic tract to the suprachiasmatic nuclei (SCN) in the anterior hypothalamus. The next link is paraventricular nuclei from which the signal first goes to intermediolateral cell column of the spinal cord and then to noradrenergic neurons in the superior cervical ganglia which provide a sympathetic innervation of epiphysis [27,29]. The binding of noradrenalin to β_1-adrenergic receptors of pinealocytes leads to adenylate cyclase activation, increased intra-cellular concentration of $3'$-$5'$-cyclic adenosine monophosphate (cAMP), stimulation of the NAT activity, and subsequently to higher MT production [407,548]. In pinealocytes, there are also α_1-adrenoreceptors which, when stimulated by noradrenalin, potentiate the activity of β_1-adrenoreceptors via Ca^{2+}–phospholipid-dependent protein kinase C. The simultaneous stimulation of α and β adrenore-ceptors by endogenous noradrenalin may ensure a more effective potentiation of the NAT activity at a relatively low level of catecho-lamines [21,464].

Until recently it has been believed that the pineal gland is a major location of the MT synthesis in the organism. However, comprehensive studies (biochemical, immunohistochemical, radioimmunological) have found numerous extrapineal sources of this hormone production in other organs, tissues, and cells which have necessary enzymatic appara-tus (Table 2.1). Thus, it has been shown that MT is generated in the retina, lens, ovary, bone marrow, enterochromaffin cells of the intesti-nal tract, in the vascular endothelium, as well as in lymphocytes, macrophages, and platelets [1,209,240,295,369]. It has been established that MT-producing cells form a part of the organism's neuroimmu-noendocrine system [288].

The MT synthesis in the APUD cells, which are located in almost all organs and tissues, obviously provides the necessary hormone pool that helps regulate the physiological functions of vital organs *in situ* at the molecular, subcellular, cellular, tissue, and organ levels.

MT passes into the capillary blood by passive diffusion and proba-bly by forming a complex secretory product [262]. Due to its high lipid solubility, MT penetrates into all biological fluids and each cell of the organism [105]. None of the existing morphophysiological bar-riers (e.g., the histohematic barrier) can prevent the MT passage [475]. Consequently, MT has a unique impact on every cell of the human body. In spite of the fact that 70% of MT circulating in blood is linked with albumin, large quantities of MT has been found in cerebrospinal fluid, in the ocular anterior chamber, cochlea, saliva, bile, milk, preovulatory follicle fluid, in seminal and amniotic fluids, into which, according to many researchers, it passes by diffusion from plasma [79,141,224,273,314,466].

Table 2.1 Extrapineal Sources of MT	
Organs and Tissues	**Cell Types**
Gut mucosa	EC-cells
Airway epithelium	EC-cells
Prostate, kidneys	EC-cells
Bladder, urethra	M-cells
Uterus, ovaries, placenta	Apudocytes
Thyroid	C-cells
	B-cells
Liver	M-cells
Gallbladder	EC-cells
Thymus	Apudocytes
Adrenal glands	Apudocytes
Paraganglia	Chromaffin and nonchromaffin apudocytes
Pancreas	B-cells
Carotid body	M-cells
Cerebellum	M-cells
Retina	M-cells
	Endothelial cells
	Platelets
	Eosinophils
	Mast cells
	NKs
Source: *According to Raikhlin and Kvetnoy [458].*	

The MT half-life in blood is rather short—10–40 min [21]. Circulating MT is metabolized mainly (90%) in the liver where around 75% of the hormone under the influence of microsomal enzymes transforms into 6-hydroxymelatonin [689]. Hydroxylated metabolites of MT are excreted with urine in form of sulfates (70%) and glucuronides (6%) (Figure 2.2). In the brain, around 12% of MT first transforms into N-acetyl-N-formyl-5-methoxykynuramine, and then into N-acetyl-5-methoxykynuramine, which is structurally similar to ASA. It is worth noting that antioxidant and anti-inflammatory properties of MT are mostly associated with the formation of these metabolites, of which the latter is the strongest [256,348,478,576]. Also, in different organs other metabolites of MT can form, and around 1% of MT does not metabolize at all. All MT metabolites are excreted with urine [306].

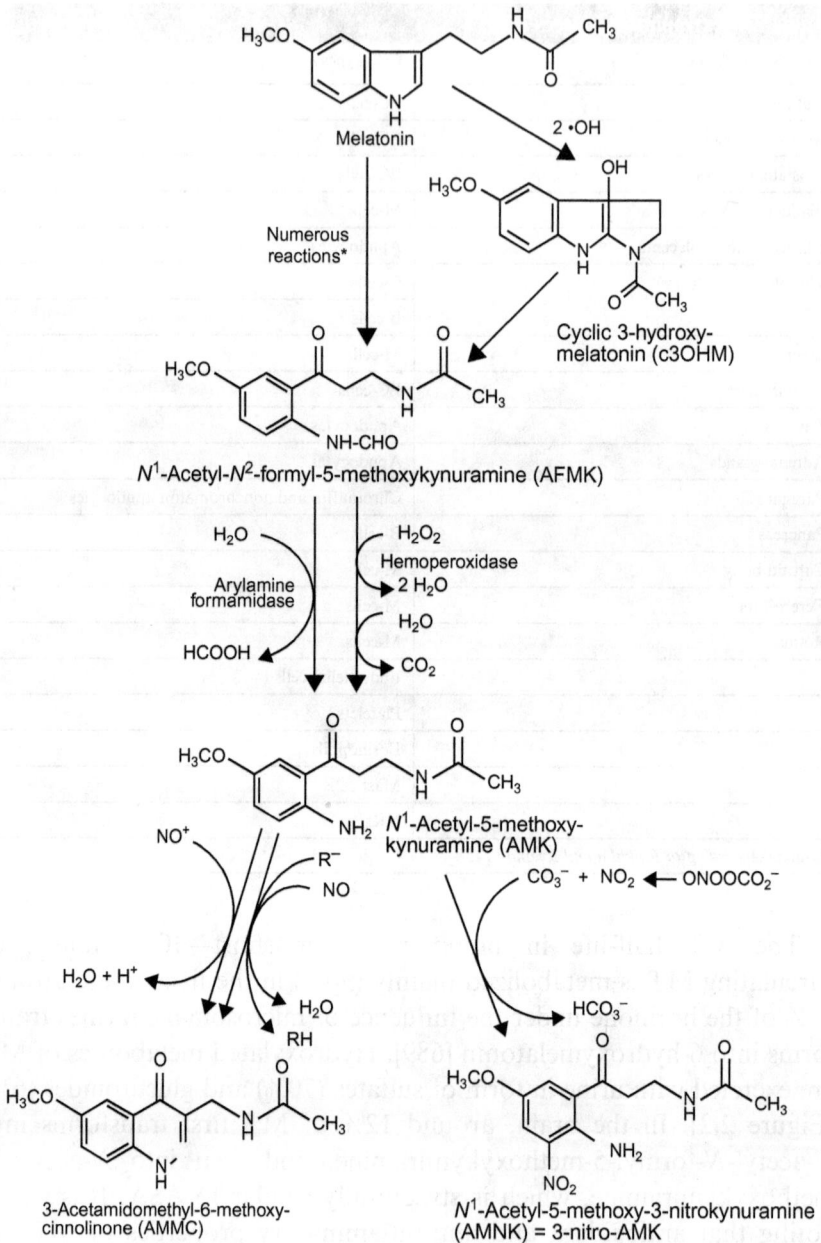

Figure 2.2 The kynuric pathway of MT metabolism, including recently discovered metabolites formed by interaction of N^1-acetyl-5-methoxykynuramine (AMK) with reactive nitrogen species. From Pandi-Perumal et al. [407]

2.2.2 The Role of MT in Regulation of Biological Rhythms

It is known that during MT production, there exists an endogenous rhythm of its substantial increase at night due to the increased synthesis of the hormone in the pineal gland, which begins at 9–10 p.m., reaches its maximum at around 2–4 a.m., and decreases by 7–9 a.m. [21]. It has been shown that this process is connected with changes in the activity of SCN and NAT [27,267,483]. The decrease of the NAT activity, and therefore of the MT production, occurs in the late dark period and is a consequence of ceasing the NAT transcription, thanks to the actions of a specific protein [543].

The endogenous daily rhythm of the MT production is greatly affected by the impact of light. The activation of retina photoreceptors suppresses the SCN activity and causes a rapid fall of the MT production and, consequently, its level in circulation [305]. Therefore, at daytime the MT level in plasma remains rather low. At that time, extrapineal sources play a substantial role in maintaining the MT level in plasma and tissues. This is evidenced by the fact that in pinealectomized rats, MT has been found constantly although at a relatively low level [400].

The brighter the light, the greater the suppression of the MT synthesis [67]. In absence of light (e.g., in blind humans), the MT secretion rhythm becomes free-flowing and seems to reflect the endogenous activity of hypothalamus SCN, the period of which lasts approximately 24 h [192]. Thus, light is necessary for synchronizing the biological rhythms in humans, including the rhythms of MT secretion, with the rhythms of the environment.

The MT synthesis in humans also has a seasonal rhythm. This may be connected with the increased MT production at night in proportion to the dark-phase duration [474]. Therefore, in autumn months, as the dark phase becomes longer, the hormone synthesis increases, whereas in spring, as the daylight hours lengthen, the situation is reversed [472,584].

Besides the seasonal and circadian rhythms, an ultradian rhythm of MT has been revealed. Thus, in prepubertal children, 3.1–3.4 hormone secretion peaks occur every 4 h [99], and in adults 4.0–4.5 peaks occur every 12 h [421]. At night, the rate of MT secretory fluctuations increases 10 times on average [193].

It has been established that, besides the rhythmic MT production, its synthesis undergoes regular changes connected with age periods in human life. Thus, in newborn infants the MT production is low, but it increases rapidly, reaching its maximum by 3–5 years, and after 5 years there is a substantial decrease in the MT synthesis; in childhood it becomes the same as in adults [21,477]. Then, after 40, the MT production gradually goes down with aging and reaches a very low level in the old-age group [473]. At that period, the daily rhythm of the hormone secretion disappears, which is thought to be linked with the activity of enzymes involved in the MT production or a decrease of the quantity of β-adrenoreceptors on the pinealocytes membranes in elderly humans [317,585].

As a central endogenic synchronizer of biological rhythms, MT exerts its influence at the organ, cellular and subcellular levels by binding with its own receptors on the membranes of target organs. There are two types of receptors for MT with high and low affinity: MT1 and MT2 [185,409]. These receptors belong to the G-protein-coupled receptor family, and their activation leads to adenylate cyclase inhibition [291,303,513,604]. The MT1 receptor is also involved in phospholipase C activation, and the MT2 receptor also mediates the guanylate cyclase inhibition [610]. The MT1 receptors activation modulates neuronal induction, mediates arterial vasoconstriction, tumor cells proliferation, reproductive and metabolic functions of the organism. The activation of MT2 MT receptors phase shifts circadian rhythms of neuronal firing in the SCN, inhibits dopamine release in the retina, induces vasodilatation, inhibits leukocytes rolling in the arterial beds, and enhances the immune response [132,138].

The third binding site for MT has been discovered recently. Newly-purified MT3 protein belongs to the family of the quinone reductases. The inhibition of the quinone reductases-2 by MT is connected with the well-known antioxidant properties of this hormone. The MT3 receptor is also involved in the regulation of intraocular pressure [389,439]. There is another receptor (GPR50) which is not MT binding but acts as MT1 receptor antagonist [245].

It is worth noting that both the central and peripheral MT receptors gave a periodic sensitivity to the hormone circulating in blood. In refractory periods, the receptors do not respond to MT. Therefore, when the rhythms of MT production and "sensitivity windows" do not coincide, even high concentrations of the hormone over a long time cannot impact the function of an organ [465,473]. Receptors to MT have been discovered in many organs and tissues: in different divisions of the CNS, blood vessels, heart, breast cells, gastrointestinal tract, liver, gallbladder, kidneys, uterus, testes, prostate, skin, adipocytes, and immune system cells [251,408]. It has been shown that MT can bind with nuclear receptors that regulate gene expression [80,413,534,592]. Furthermore, the hormone may directly impact the cell by combining with calmodulin [236].

The existence of specialized receptors to MT in different organs and tissues, as well as the hormone's ability to cross morphophysiological barriers and reach all intracellular structures, allows it to take part in homeostatic regulation, ensuring the organism's adaptation to various exogenous and endogenous conditions. By providing the endocrine support to the rhythm-setting function of the central pacemaker, MT influences the circadian dynamics of various physiological data [30]. Thus, MT enhances the decline of body temperature at night and acts as a natural inductor of physiological sleep [5,76,192]. The administration of MT to blind people at 11 p.m. helps synchronize the impaired "sleep-wake" cycle by suppressing the reaction that results in them falling asleep during the daytime [492]. Seasonal fluctuations of the MT production underlie the seasonal alterations in humans, and pineal deficiency is manifested in disadaptive disorders underlying the seasonal recurrence of chronic diseases. This is confirmed by the absence of seasonal cycling of the MT production in patients with malignancies, as well as by more frequent depressions and alcoholism in people with impaired seasonal rhythms of MT secretion due to relocating from middle latitudes to the far north [21,46,584].

2.2.3 MT as an Endogenous Antioxidant

In recent years, researchers have focused on another extremely important function of MT—its participation in the regulation of the oxidation–reduction process in the organism. It is known that lipid peroxidation (LPO) occurs in all cells and tissues as a normal metabolic process that provides renovation of lipids and reconstruction of membrane structures. The LPO activation in different pathological conditions may damage the cell adaptation capabilities [555]. In such conditions, MT plays an important role as a powerful endogenous antioxidant. Being an electroactive molecule, MT acts as an electron donor, thereby detoxifying the oxygen-reactive radicals, such as hydroxyl radical ($\bullet$OH), peroxyl radical (LOO$\bullet$), superoxide anion radical ($O_2^{-\bullet}$), and peroxynitrite anion (ONOO$^-$) [197,326,410,468]. MT can also scavenge hypochlorous acid (HOCl), which is generated by activated monocytes and macrophages in reactions of bacteria uptake and elimination, and directly scavenge the gaseous neurotransmitter— nitric oxide (NO$\bullet$) [388,466]. The latter can also be highly toxic when it is excessively released during ischemia−reperfusion injury [211,408,429,609]. By giving out its electrons, MT transforms in various metabolites (N-acetyl-N-formyl-5-metoxykynuramine and cyclic 3-hydroxymelatonin), which do not any more participate in oxidation−reduction reactions as other antioxidants and thereby stop the process of oxidation and free radicals formation [577]. Furthermore, it has been shown that MT indirectly affects the neutralization of singlet oxygen ($^1O_2$) and the activation of antioxidant defense enzymes (superoxide dismutase, glutathione peroxidase, glutathione reductase, glucose-6-phosphate dehydrogenase) [401,477,644]. MT, as a natural antioxidant, has an antiapoptotic action on healthy cells [77,116,450].

2.2.4 Role of MT in Regulation of Platelet-Vessel Hemostasis

The state of platelet-vessel hemostasis is known to determine the development of pathological processes in man, and particularly, the disorders of microcirculation and hemodynamic [417,427]. As a consequence, the role of MT in regulating different parts of the hemostasis system has been attracting researchers' attention in recent years [606,640]. It should be noted that most studies focus on the impact of MT on the functional activity of platelets as the main reservoir of serotonin and, consequently, a source of MT production in man [258,368,580]. Not only does the MT synthesis occurs in platelets, but MT transforms in other metabolites as well. Thus, when using exogenous MT, its metabolite 5-sulphatoxy-N-acetyl-serotonin (SNAS) is formed in platelets [123,127].

In turn, MT has a substantial effect on the functional activity of platelets. It has been proved that MT in large doses (2×10^{-8} M and 2×10^{-3} M) inhibits serotonin uptake by platelets, while physiological concentrations of the hormone (10^{-12} to 2×10^{-9} M) do not influence either uptake or spontaneous release of 5-HT by platelets. At the same time, addition of serotonin (10^{-6} M) inhibits the binding of MT by blood platelets [339,362,602]. It has been shown that the highest sensitivity of platelets to MT-inhibiting effect on serotonin transport exists in the morning (8 a.m.), while the release of 5-HT from platelets is not time dependent [339,361,362]. Other researchers have demonstrated the most manifest nocturnal (10.30 p.m.) inhibiting effect of MT ($10^{-9}-10^{-5}$ mol L^{-1}) on release of (3H)-5-HT induced by thrombin in washed human platelets [117]. Various effects of MT on human platelets are probably mediated by MT binding to highly sensitive acceptors which have been found on the membrane of human platelets in experiments with labeled (3H)-MT [517,600,639].

It is known that platelets' functional activity is, to a certain extent, predetermined by biochemical activity of precursor cells of thrombocytopoiesis—megakaryocytes [538]. The depth of structural disorders in megakaryocytes seems to determine platelets dysfunction in different pathological states, which is expressed in impaired aggregation, release reaction, and blood clot retraction. The studies of MT effect on megakaryocytes have demonstrated that MT easily crosses the megakaryocyte membrane and accumulates within the nucleus [122]. MT interferes with some phases of translation and transduction, exerting β-cytochalasin-like action, that is, inhibits the process of endoreduplication and increases the nucleus ploidy [487]. MT can accelerate the formation and development of mature megakaryocytes [121] and increase the intensity of platelet formation [123,488]. The authors have shown that the addition of MT to a suspension of bone marrow cells leads to the thinning of the megakaryocytes' membrane, their destruction, and finally to the formation of a platelets cluster. This process resembles the formation of cytoplasmic fragments from the body of megakaryocytes which then flow through gaps of the sinusoidal membrane and turn into platelets going to the bloodstream [487]. MT contained in platelets' cytosol partly deacetylases into serotonin, and remaining MT stabilizes the platelet membrane, preventing the aggregation of blood platelets and extending their life [123].

As we know, platelet functional activity is manifest in adhesion and aggregation. Numerous *in vivo* experiments have demonstrated the effect of MT on almost all phases of platelets activation. Thus, by binding to the microtubule membrane of platelets, MT induces the change of the platelet form from a flat disk to sphere [487]. MT inhibits the platelets aggregation induced by collagen, arachidonic acid (AA), adenosine diphosphate (ADP), adrenalin, thrombin, calcium ionophore A23187 [117,153,274], as well as the ADP-induced ATP release and formation of cyclooxygenase (COX) products of AA metabolism [78,117]. All these effects of MT are dependent not only on the hormone concentration but on the time of day and night. Thus, MT in concentration of $10^{-7}-10^{-5}$ M L^{-1} inhibits ADP-induced aggregation of platelets at night only, while the suppressing effect of MT ($10^{-9}-10^{-5}$ M L^{-1}) on ADP-induced ATP release is dose dependent and occurs during the whole day, although mostly in dark hours (10.30 p.m.) as compared to morning hours (8.30 a.m.) [117]. A study of the MT effect on platelets aggregation induced by AA has proved its suppression only in blood samples which were taken at 1.30 a.m. At the same time, there has been observed a maximum decrease of thromboxane (Tx) B_2 (TxB$_2$) production under the influence of the same inductor. The minimum MT effect has been observed at 3.30 a.m., that is, in the period of the highest concentration of the hormone in human plasma [601]. Therefore, there are not only daytime variations of platelets sensitivity to MT but nocturnal variation as well. Along with this, researchers have observed various effects of MT on COX metabolites of AA. Thus, MT in micromole concentration inhibits the transformation of AA into PGF$_2$ and TxB$_2$ to a greater extent than into PGE$_2$ and PGD$_2$ [342]. It is known that in physiological conditions, the platelets activation leads to formation of three major eicosanoids: Tx A$_2$, 12(S)-hydroxy-5, 8, 10-heptadecatrien acid, which forms from intermediate instable endoperoxide PGH$_2$, and 12-hydroxyeicosatetraenoic acid (12-HETE). The latter is generated from AA by the action of 12-lipoxygenase (12-LO) [320]. As MT in epiphysis reduces the expression of messenger RNA (mRNA) 12-LO and the level 12-HETE in rats, we may assume a similar effect of the hormone on the 12-LO activity in platelets.

Furthermore, the process of platelets formation from megakaryocytes, in which MT plays a leading role, is followed by the synthesis of platelet-activating factor (PAF) [488]. The latter is partly retained within the cell and partly released from it [123]. The interaction between PAF and platelets leads to their activation, aggregation, and to release of different biologically active substances (serotonin, platelet factor 4, Tx A$_2$, etc.) [233]. Numerous studies have demonstrated yet another action of PAF—an increase of vascular permeability [143].

The membrane receptor complex is a universal primary regulating system which to a large extent depends on membrane lipid microviscosity [598]. An increase of the rigidity of membrane phospholipid structures due to insufficient function of the antioxidant system (AOS) and accumulating LPO products leads to a change in surface receptor's exposure. It affects the platelets' reactivity to an inducing agent [242,355,462]. Having antioxidant properties, MT, on the one hand, reduces the ability of free radicals to damage the deoxyribonucleic acid (DNA) structure [467,469], and on the other hand, it substantially increases fluidity and decreases microviscosity of platelet membranes [114]. By easily penetrating the lipid bilayer of cell membranes, MT protects phospholipids from free-radical damage and consequently from platelet apoptosis [105,517]. Therefore, by changing the functional activity of platelets, MT plays an important role in hemostasis regulation.

An important characteristic of the vascular functional status is thromboresistance of vessels that is the formation and release by vascular wall cells of thrombogenic and athrombogenic substances [427]. Regrettably, in literature we have not come across any data on a direct influence of MT on that process. However, in recent years there have started to appear more information about the MT involvement in the regulation of vascular tone. Thus, MT induces a dose-dependent relaxation of pulmonary arteries and veins, and its effect is mostly expressed on the artery's vascular smooth muscles [623,624]. Experimental studies have demonstrated the vasorelaxing action of MT on isolated aorta and basilar artery in rabbits and rats [504,525,631]. In some instances this effect is linked with an increase of 3'-5'-cyclic guanosine monophosphate (cGMP) level in cells; in other instances it is related to inhibition of calcium channels or the hormone's influence on perivascular nerve endings [623]. As for coronary vessels, MT, in contrast, causes their dose-dependent reversible involution [622]. Such effect of the hormone has been demonstrated in respect of caudal arteries and small cerebral arteries in rats [73,144,463]. The varied influence of MT on vessels may be due to the hormone's effect on different subtypes of smooth muscles receptors. Thus, the activation of MT2 MT receptors may induce vascular relaxation, while the effect on the other receptor subtype may cause vasoconstriction [131]. MT may have not only direct but mediated influence on vascular tone. Thus, MT inhibits the noradrenalin response in the middle cerebral artery of ape's fetus, thereby preventing its involution [583].

Interesting data can be found in the article by Monroe and Watts [367] on the simulating effect of MT on the vascular reactivity which mostly is of nonspecific nature. Additionally, there are seasonal variations of vascular reactivity that are connected with seasonal rhythms of the hormone secretion [110].

It is known that enhancement of LPO and free radicals formation may greatly affect the vascular tone. By neutralizing almost all oxygen-reactive radicals, MT substantially inhibits the vasospastic action of H_2O_2 [396] and prevents DNA fragmentation and apoptosis of cerebral endothelial cells [523].

MT has been proved to have an inhibiting effect on edema formation and leukocytes adhesion to the venule wall after ischemia and reperfusion [54,611]. Some researchers have noted a mediated action of MT on vascular tone by way of reducing the activity of NO-synthase (NOS) in endothelial cells, which may play an important role in preventing a hypotensive effect connected with cytokine therapy of cancer patients [311,322,323]. Other researchers believe that the protective effect of MT on microcirculation after prolonged ischemia is connected with its antioxidant action [614]. MT can also contribute to reducing the vascular permeability, which is connected with decreased production of NO and vascular endothelial growth factor [254]. In addition, MT is able to suppress proliferation of human umbilical vein endothelial cells by inhibiting the expression of NF-κB and activating various intracellular signaling pathways [109].

Thus, there is no doubt that MT is directly and indirectly involved in the regulation of platelet-vessel hemostasis, although many aspects of this problem still need to be studied further. In particular, special attention should be paid to MT's effect on NO, a universal regulatory molecule involved in regulating function state of platelets and vascular tone.

2.2.5 MT's Effect on Nitric Oxide Production in Organism

NO is a highly reactive molecule which constantly forms in live organisms and acts as an important regulator of physiological functions. It is known that a source of the endogenic synthesis NO in cells of mammals is amino acid L-arginine [404]. NO is synthesized from arginine by the enzyme NOS. This reaction occurs in the presence of oxygen, but in the case of its deficiency, the "time-tested phylogenetically old mechanisms of adaptation" may become active [90]. They lead to the formation of NO from its metabolites: ions NO_2^- and NO_3^-. This method of NO synthesis is more economic in terms of energy used, and it is synthesized by means of nitrite reductase systems connected with heme-containing proteins—hemoglobin, myoglobin, cytochrome oxidase and cytochrome P-450 [482]. Oxygen deficiency serves as a signal that triggers transition of cell to nitrate/nitrite breathing, which activates the system of xanthine oxidase–xanthine dehydrogenase, ions NO_3^- restore to NO_2^-, heme-containing proteins transform into desoxy form, and ions NO_2^- begin to actively restore to NO by accepting electrons from these heme-containing proteins. The existence of NOS and nitrite reductase systems creates conditions for forming a closed cycle which was called the "NO cycle" by Reutov et al. [480]. This provides a continuous production of NO in mammals, not only in the case of oxygen sufficiency, but also in case of its deficiency, which occurs during hypoxia, functional load, and pathological processes (Figure 2.3). This is especially important because NO performs an important function of regulating intra- and intercellular processes. Primarily, NO participates in regulating the intercellular concentration of ions Ca^{2+} [221] and acts as super antioxidant within a wide range. Thus, in presence of O_2, O_2^-, and H_2O_2, NO transforms to less active compounds—ions NO_2^- and NO_3^-, which may excrete from

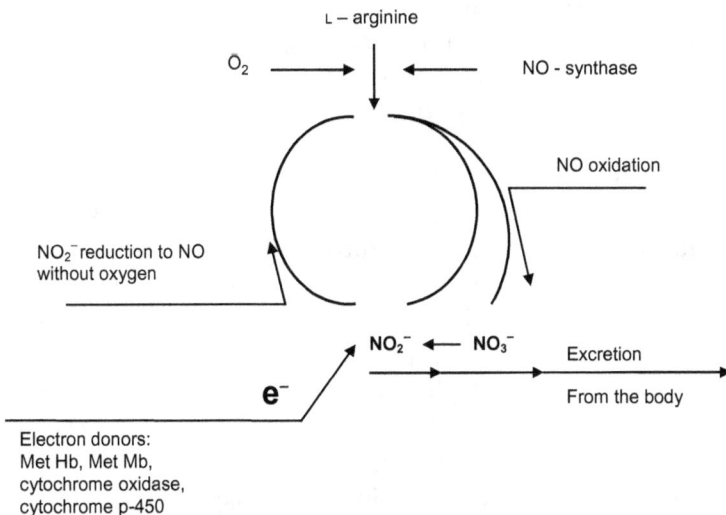

Figure 2.3 NO cycle in organisms. From Reutov [481]

mammals' organisms through the kidneys [481]. Besides, the activation of the "NO cycle" and increase of NO and NO_2^- may induce the transition of proteins from a soluble to membrane-bound state, in which many enzyme systems are activated, including those involved in the synthesis of ATP [8,479]. NO acts as a signaling molecule in different neuronal functions [214,337,455], has a cytotoxic effect in mechanisms of nonspecific immune protection [282], and also acts as a tumor-inhibitory and bactericide agent in a number of autoimmune diseases [394,625]. NO participates in the relaxation of blood vessels and hyperplasia of muscle cells of their wall [68,137,200] and performs many other important functions in the human organism [597,364].

The synthesis of NO from L-arginine is catalyzed by a family of enzymes referred to as NOSs [269,522]. These enzymes are subdivided into two major groups—constitutive (cNOS) and inducible (iNOS). The constitutive enzymes are constantly expressed in cells and regulated by calmodulin and physiological concentrations of Ca^{2+} in the cell [48]. The cNOS plays a "physiological" role and participates in providing intercellular interaction and intracellular reactions [393]. The iNOS, which is activated some time after the impact of cytokines and bacterial lipopolysaccharides, is closely connected with calmodulin but is not dependent on physiological concentrations of Ca^{2+} in the cell and produces large quantities of NO [374,453]. The expression of iNOS is also induced by interleukin 1β (IL-1β), tumor necrosis factor (TNF-α), interferon-gamma (IFN-γ); the iNOS is activated by plasmin, fibroblast growth factor (FGF), and inhibited by insulin-like growth factor (IGF), platelet-derived growth factor (PDGF), transforming growth factor β, and thrombin [74].

Currently, three isoforms of NOS have been extracted, which differ by their location in different cells and their own gene coding: neuronal, endothelial, and inducible [481]. The neuronal NO-synthase (nNOS), which was initially extracted from neurons of the CNS, is a constitutive cytosolic enzyme [393]. Experimental studies have shown that nNOS is expressed in the bronchial and intestinal epithelium, in human skeleton muscles, endothelium, and photoreceptors [510]. The endothelial NO-synthase (eNOS) is found in the endothelium of blood vessels, endocardium, myocardium and also is constitutive [241]. The inducible form is denoted as iNOS and expressed in macrophages, cardiac histiocytes, vascular smooth muscle elements, intestinal epithelium, megakaryocytes, and hepatocytes [7,352,390]. It should be noted that the three forms of NOS (nNOS, eNOS, and iNOS) are also expressed by platelets [454]. It has been established that the human nNOS gene is located in the 12th chromosome; the eNOS gene, in the 7th one; and that of the iNOS, in the 17th chromosome [269].

In recent years, researchers and clinicians' attention has been drawn to studying the role of NO in the pathogenesis of lung diseases. An increased level of NO in the expired air has been found in patients with bronchial asthma [261,344], bronchiectasis [260], seasonal rhinitis [341], and upper airways infections [261]. The increased NO production in patients with such diseases is probably connected with enhancing the expression of iNOS in the upper and lower airways [190,213,308,318,508] and may act as a marker of inflammation in the bronchopulmonary apparatus. At the initial stage, the higher synthesis of NO may be a compensatory adaptive reaction of the organism to hypoxia and it helps decrease the activity of inflammatory cells, kill microorganisms, improve blood circulation, ciliary transport, and bronchodilatation [45,51]. But such excessive accumulation of NO may have an opposite adverse effect due to the formation of nitrogen dioxide, peroxynitrite, and hydrogen peroxide, which can cause oxidative damage to vascular endothelium and thereby enhance vascular permeability and formation of an inflammatory edema [302,454]. Furthermore, the high concentration of NO in epithelial cells may suppress the cNOS activity, exhausting the pool of soluble guanylate cyclase, thereby increasing the intracellular content of Ca^{2+} and contribute to airways constriction [382]. In case of damaged epithelium, the excessive synthesis of NO strengthens inflammation and vagus' constrictory effects [380]. Thus, the activation of the NO cycle during a moderate hypoxia may cause changes similar to mobilization (activation of processes), and then, as NO and ions NO_2^-, NO_3^- accumulate and the proportion between cAMP and cGMP changes to an increase of the latter, there occurs a protective suppression connected with inhibition of intracellular processes [481]. Such regularity of compensatory adaptive reactions can be defined by the notion of nonspecific adaptive syndrome and presupposes the existence of "stress-limiting systems" which directly protect cellular membranes from damage [349]. Therefore, it is important to look at the modulating influence of MT on the "NO cycle." It has been established that MT as a powerful natural antioxidant regulates the NOS activity and neutralizes free radicals which form in the process of NO synthesis and metabolism [36,112]. It is known that NO is synthesized in epiphysis, and its production increases under the influence of noradrenalin which stimulates the Ca^{2+} calmodulin-sensitive form of NOS [309]. There is a certain relationship between NO and MT: NO can suppress MT synthesis in epiphysis stimulated by catecholamine [338], and MT, in its turn, inhibits the activity of NOS. This effect of MT has been demonstrated in respect of both the constitutive and inducible isoforms of NOS. In the first instance, MT changes the binding of cNOS to calmodulin [446,447], and in the second instance it inhibits the enzyme expression, suppressing its transcription in part by suppressing the activated transcription factor NF-κB [55,195]. Studies of MT's effect on NO production by endothelial cells under the impact of bradykinin, carbachol, and histamine have proved its inhibiting action mediated by MT association with G-protein-coupled receptors and suppression of Ca^{2+} mobilization from intracellular stores [529]. Guerrero et al. [210] have demonstrated a reverse correlation between the nocturnal content of NO and the MT in the chicken brain. Exposure to light at night leads to decrease of the MT synthesis and increase of NO and cGMP, which the authors attribute to absence of MT's controlling effect on the NOS activity in these conditions. Other researchers, however, have shown a stimulating influence of MT on NO production in human monocytes [369]. Using NO, MT produces an oncostatic action on tumor cells in fibromatosis, at the same time regulating the cNOS [363]. In addition to the modulating influence on the NOS activity, MT exhibits antioxidant properties in respect of the NO molecule. Thus, MT and its precursors restrict oxidative damage of the brain and kidneys at ischemia and postischemic reperfusion by neutralizing NO and inhibiting its synthesis [120,211,387]. The neutralizing influence of MT has been also proved in respect of peroxynitrite, a cytotoxic radical which forms in interaction of NO with superoxide anion [197,395]. Thus, by affecting cyclic transformations of NO in the organism, MT participates in regulating the concentration of this compound and therefore coordinates intercellular relationships in the organism.

2.2.6 The Role of MT in the Regulation of Arachidonic Acid (AA) Metabolism

Metabolites of AA (eicosanoids) constitute a large group of neuroimmunoendocrine signaling molecules. AA is present in phospholipids of cellular membranes and makes up about 1% of free fatty acids in plasma, circulating as an albumin complex. When cells are activated by a stimulus that is able to change the types and geometrical arrangement of phospholipids and to activate phospholipase A_2, AA is released followed by metabolism on a COX or LO pathway [304].

In properly-functioning cells, the role of such stimulus is played by products of free-radical lipid oxidation.

The formation of prostaglandins (PGs) and Txs on the COX way goes through the stages of unstable biologically inactive PGG_2 and PGH_2. Then COX metabolites are synthesized in different cells in different ways, depending on the enzyme that is prevailing in those cells [304,414]. The COX enzyme PGH-synthase has two isoforms—cyclooxygenase 1 (COX-1) and cyclooxygenase 2 (COX-2), which have 61% of one-type sequence of amino acids. COX-1 and COX-2 mediate physiological and inflammatory processes respectively and react to different stimuli by formation of prostanoids [581]. COX-1 is present in platelets, endothelial cells, gastric mucosa, kidneys, etc., while COX-2 is synthesized *de novo*, mostly in macrophages but also in lungs, heart, vessels, and spleen. It is responsible for massive noncontrolled formation of prostanoids in case of stimulation by bacterial endotoxins and cytokines [268].

The LO pathway of AA metabolism leads to the formation of various LTs, mono-HETEs, and lipoxins (LXs), whose synthesis, like that of COX products, depends on an enzyme prevailing in cells (Table 2.2). LXs (A and B) are trihydroxy acids which are generated from AA as a result of consecutive action of two LOs: 15-LO and 5-LO. LXs serve as a sort of "stop signals" regulating key aspects of leukocytes transport and preventing the development of acute tissue damage mediated by those cells [83,84,519−521,537]. The enzyme 5-LO has been found only in myeloid lineage cells.

Epithelial cells of human airways, eosinophils, mast cells, and dendritic cells contain a large quantity of the first isoform 15-LO-1. The second isoform of this enzyme (15-LO-2) also exits in these cells. Under the action of 15-LO enzyme in eosinophils, not only LTs C_4, D_4, E_4 are formed, but also metabolites that have similar chemical structure ($14,15\text{-LTC}_4$; $14,15\text{-LTD}_4$; $14,15\text{-LTE}_4$) and are called eoxins (EX) [181]. Cells that have the full set of enzymes (eosinophils, mast cells, and basophils) are able to generate large quantities of sulfidopeptide LTs (LTC_4, LTD_4, LTE_4) [581].

Table 2.2 Cellular Sources of Eicosanoids in Lungs

Eicosanoids	Cellular Sources
PGE_2	Bronchial epithelial cells, smooth muscles of central and peripheral bronchial tree, macrophages, lymphocytes
PGI_2	Vascular endothelium, vascular and nonvascular smooth muscles
PGF_2	Macrophages, smooth muscles of central and peripheral bronchial tree
PGD_2	Mast cells
TxA_2	Platelets, alveolar macrophages
LTB_4	Neutrophils, alveolar macrophages, monocytes
$LTC_4/D_4/E_4$	Direct synthesis from their own membrane: mast cells, eosinophils, basophils. Transcellular metabolism: neutrophils and platelets, neutrophils and vascular endothelium
5/12/15-HETE	Neutrophils, platelets, bronchial epithelial cells
LXs	Epithelial cells with neutrophils, mast cells or macrophages, platelets with granulocytes
Eoxins	Eosinophils, mast cells, basophils

Platelets have the enzyme LTC_4-synthetase but not 5-LO. Due to this, platelets can generate LTC_4 only from LTA_4, formed by neutrophils through the mechanism of transcellular metabolism. In the same manner, platelets generate LXs, this process being catalyzed by platelet 12-LO [134,485,521]. A similar mechanism of intercellular relationship exists between neutrophils and vascular endothelial cells. The biosynthesis of LT requires the transmembrane protein known as 5-LO-activating protein which plays a certain role in binding 5-LO to phospholipids of cellular membranes for initiating the catalysis [581].

In properly-functioning cells, the hydrolysis of membrane lipids with release of AA occurs at a low level, which provides a small quantity of eicosanoids. In physiological conditions, there also exist systems that inhibit their synthesis. The inhibiting effect is produced, in particular, by lipocortin, a highly polar protein that is present in different cells including monocytes and neutrophils. The formation of lipocortin is regulated by the level of circulating corticosteroids which induce its formation. The effect of lipocortin is associated with inhibition of phospholipase A_2, which suppresses the release of AA from phospholipids and thereby blocks the formation of PGs, LTs, and PAF. The activity of COX and LO is regulated by fatty acids hydroperoxides, which, even in small quantities, activate these enzymes. The pathologic signal increases according to a "vicious circle" mechanism. The mechanism of a system that returns to the physiological level is probably associated with autocatalysis and autoinhibition of enzymes, for reproduction of which a certain time is required [293,414].

In lungs, there exists a system of controlling the formation and inactivation of eicosanoids. It has been established that bradykinin, histamine, and LTs increase the synthesis of $PGF_{2\alpha}$ and accelerate its release in allergic reactions, and LTC_4 increases the synthesis and excretion of TxA_2 [188]. PGE_1, PGE_2, and PGI_2, in their turn, when added to macrophage homogenates, enhance the adenylate cyclase activity, increase the synthesis of cAMP and thereby decrease the production of LTs, particularly in basophils [418]. It has been established that PGI_2 also decreases the sensitivity of bronchi to bronchoconstriction effect of LTC_4 and LTD_4 [205,304,646]. Potent inhibitors of LT synthesis lead to a simultaneous decrease of 5-HETE, LTB_4, LTC_4 formation and an increase of PGD_2 synthesis. Therefore, exclusion of one enzyme pathway in AA metabolism may have an integrated effect on the formation and excretion of other mediators of inflammation [167,232].

The pulmonary system of PG inactivation consists of a transporter and intracellular enzymes. The leading role in PG inactivation belongs to the microvascular endothelium. PG oxidation results in the formation of inactive metabolites which are washed away from the lungs and undergo further transformations in the liver. Interestingly, the lungs metabolize not only the PG circulating in blood, but also the PGs that are synthesized in the lungs themselves, which should be viewed as a defense reaction resisting the admission of excessive PG to arterial blood [555].

Oxidation is a leading mechanism of eicosanoids catabolism. After immunologic and nonimmunologic stimulation, basophils, neutrophils, and eosinophils are able to generate toxic oxygen radicals which form hydrogen peroxide in the subsequent reaction $(O_2^- + O_2^- + 2H^+ \rightarrow O_2 + H_2O_2)$. At high concentrations, hydrogen peroxide inactivates sulfidopeptide LTs (LTC_4, LTD_4, LTE_4), and at low concentrations together with eosinophil peroxidase and myeloperoxidase of neutrophils and monocytes, it oxidized and destroys LTB_4, LTC_4, LTD_4, LTE_4, PGE_2, and 6-keto-$PGF_{1\alpha}$. LTs' degradation products are their biologically inactive metabolites (Figure 2.4). Another way of LTs' degradation is the production of OH• radicals by activated phagocytes with the help of acetaldehyde xanthinoxidase $(O_2^- + H_2O_2 \rightarrow O_2• + OH^- + OH•)$. The OH• radicals take part in

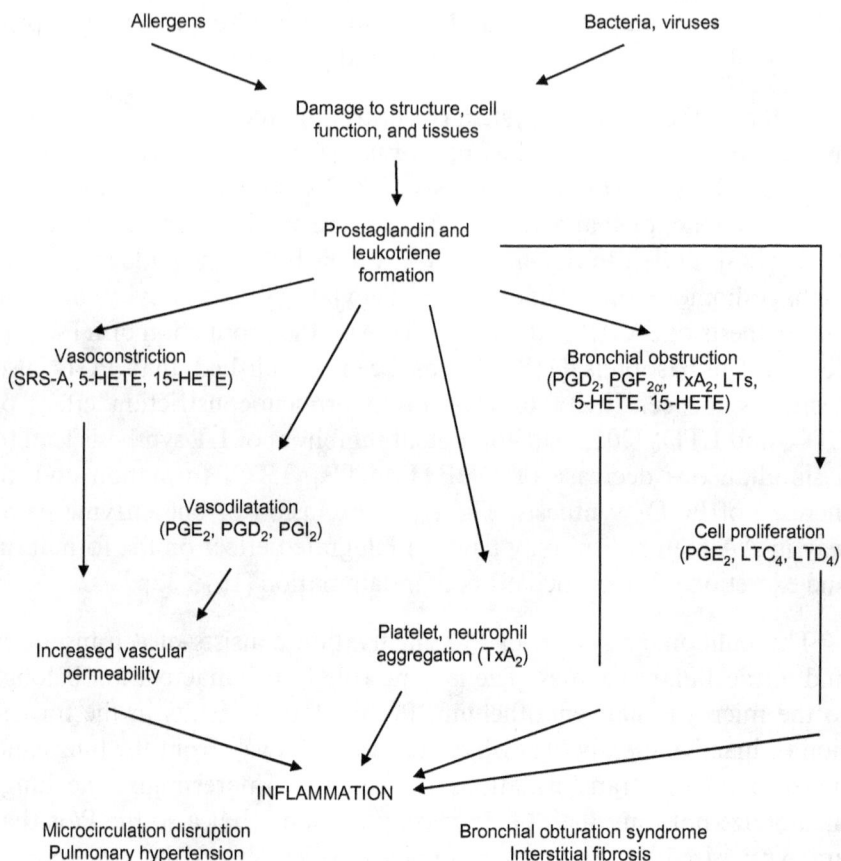

Figure 2.4 Role of eicosanoids in pathogenesis of inflammation of lungs and bronchi. From Thien et al. [581]

LTB_4, LTC_4, LTD_4, LTE_4 cleavage [226]. Sulfidopeptide LTs can also be inactivated by enzymes of the group of sulfatester hydrolases (aryl-sulfatases A and B). They have been found in LT-producing cells and tissues and in eosinophil granulocytes [452]. The main biological effects of PGs and LTs are described in Table 2.3.

The biological action of PGs and LTs is produced through receptors which are located on cellular membranes of various tissues. Bronchoconstrictors PGD_2, its metabolite 9α, 11β-PGF_2, $PGF_{2\alpha}$ and TxA_2 act on the same receptor to TxA_2. The action of LTB_4 is mediated by a specific receptor to this LT, and for all sulfidopeptide LTs there exists only one receptor to LTD_4 [581]. The receptor to LX A_4 belongs to the family of chemotaxic receptors; it is identified as

Table 2.3 Main Biological Effects of Eicosanoids

Eicosanoids	Biological Effect
PGE_1	Bronchodilatation Irritation of bronchial tree mucosa Potentiation of bradykinin effect on vascular permeability Increase of humoral immune response Suppression of T-lymphocyte function
PGE_2	Vasodilatation Increase of vascular permeability Hyperalgesia Fever Bronchodilatation Participation in immune reactions
PGD_2	Systemic vasodilatation Increase of vascular permeability Vasoconstriction of pulmonary artery (central vasoconstrictor) Bronchoconstriction (distal zones of the bronchial tree) Chronotropic effect Increase of chemokinesis and chemotaxis of granulocytes, increased effect of real chemotaxic factors Increase of histamine release by basophils
PGI_2 (prostacyclin)	Vasodilatation Increase of vascular permeability Hyperalgesia Bronchodilatation Prevention of histamine bronchospasm Inhibition of platelets aggregation Antichemotaxis effect
$PGF_{2\alpha}$	Vasoconstriction Bronchoconstriction (proximal zones of the bronchial tree) Inhibition of bradykinin, histamine, serotonin, dextran effect on vascular permeability
TxA_2	Activation and aggregation of platelets Aggregation of neutrophils Vasoconstriction Bronchoconstriction
LTB_4	Potent chemotaxis agent for neutrophils, eosinophils and monocytes Inducing adhesion of leukocytes to endothelial cells, their aggregation, and degranulation Facilitating leukocytes accumulation in tissues Facilitating the production of oxygen radicals, lactoferrin and hydrolytic enzymes by neutrophils In synergy with other mediators increases vascular permeability, especially in presence of PGE_2 Facilitates adhesion of granulocytes to endothelial cells Induces constriction of pulmonary parenchyma (by stimulating TxA_2 release) Causes bronchospasm. Takes part in immune reactions
LTC_4, LTD_4, LTE_4	Bronchoconstriction Increases the bronchoconstriction effect of histamine Increases the mucous secretion in the bronchial tree Vasoconstriction Increases permeability of systemic vessels (due to constriction of terminal arterioles and dilatation of venules) Increases permeability of small vessels Takes part in immune reactions

(*Continued*)

Table 2.3 (Continued)	
Eicosanoids	Biological Effect
14,15-LTC$_4$ 14,15-LTD$_4$ 14,15-LTE$_4$	Increase of vascular permeability Bronchoconstriction
5-HETE 15-HETE	Vasoconstriction Activation of neutrophils and eosinophils Stimulation of eosinophils and neutrophils chemokinesis and chemotaxis Potentiation of mediators release from mast cells Increase of vascular permeability Stimulation of mucous secretion
LX A$_4$	Antagonism to bronchoconstriction, caused by LTC$_4$ Vasodilatation of small vessels Inhibition of NK-cells cytotoxicity Inhibition of eosinophils and neutrophils chemotaxis, migration of neutrophils to inflammation zones Inhibition of the synthesis of inflammatory cytokines (IL-6, IL-8) Increases formation of tissue inhibitor of metalloproteinase
LX B$_4$	Inhibition of NK-cells cytotoxicity

G-protein-coupled receptor for LO metabolites and grouped with formylpeptide receptors [88]. The interaction of various classes of ligands (lipid or peptide) with this receptor regulates the participation of neutrophils in different phases of inflammatory process [87,485]. After binding by PG and LT receptors, a change in content of cAMP and cGMP is observed in cells, as followed by certain reactions. Thus, PGF$_{2\alpha}$ and PGD$_2$ increase the production of cGMP; the latter contributes to an increase of Ca^{2+} release from the intracellular stores, which leads to bronchoconstriction. At the same time, the bronchospasm caused by LT is preconditioned not only by the increase of cGMP but also by a decrease of cAMP [646]. Sometimes a positive feedback is formed between cAMP and PGs. Thus, some PGE$_2$ activates adenylate cyclase, thereby increasing the production of cAMP which, in turn, stimulates the synthesis of PGE$_2$ down to the self-catalyzing inactivation or "suicide" of COX [637].

In physiological conditions, PGs and LTs are present in tissues as trace components. In pathology, the formation of eicosanoids changes as a result of stimulation of cellular surface by immunologic and nonimmunologic stimuli. Thus, in macrophages the synthesis of PGs increases under the impact on their membrane of different agents: *Corynebacterium parvum*, endotoxin, ionophore Ca^{2+}, antigen–antibody complexes, etc. [414]. A substantial role in activating the synthesis of PGs is played by PGF-A (prostaglandin-forming factor of anaphylaxis), which is synthesized *de novo* in the process of anaphylaxis reaction. This mediator stimulates the formation and release of $PGF_{2\alpha}$ and PGE_2 from pulmonary veins, parenchyma, and airway tissue; TxB_2—from pulmonary arteries and LTB_4—from epithelial tissue of bronchi [340]. In addition, different complement components take part in activation of eicosanoids synthesis. Thus, C3a induces the secretion of TxB_2 from macrophages, C5a stimulates the formation of LTB_4 by neutrophils from exogenic AA and LTD_4 by pulmonary tissue; C5b causes a release of AA, 6-keto-$PGF_{1\alpha}$, PGE_2, and LTC_4 from macrophages; and anaphilotoxins C5b, C7, C8, C9 also facilitate the release of TxB_2 [225,414]. The formation of $PGF_{2\alpha}$ can increase under the impact of histamine which is released from pulmonary mast cells after antigen provocation of lung fragments passively synthesized *in vitro* [225]. PAF is a potent stimulus for generation of LTs and TxA_2. PAF is found at the third stage of an anaphylactic reaction where mast cells and basophils are stimulated with the release of vasoactive amines [233]. PAF and AA are released from a common predecessor (1-alkyl-2 acyl-glycero-3-phosphocholine) in various cells. In both cases, there is a need of activating phospholipase A_2 enzyme. They are closely interrelated: PAF regulates the metabolism of AA, while AA, in its turn, regulates the synthesis of PAF. The synthesis of LTs, as well as their release from mast cells, neutrophils, basophils, and eosinophils, is induced by the direct interaction of bacterial exotoxins and thiol-activated toxins with specific receptors on polymorphic nuclear granulocytes [272].

An increased content of AA metabolites and a change of balance between them have been found in many pulmonary diseases, which has helped understand their role in the pathogenesis of bronchial asthma, pneumonia, respiratory viral infections, pulmonary hypertension, pulmonary edema, vascular thrombosis, etc.

A basis of most pulmonary and bronchial diseases is inflammation as a reaction to some local damage which may be caused not only by action of different microorganisms (bacteria, viruses, etc.), but also by substances of antigenic or haptenic nature. Allergy or hypersensitivity to allergens, which develops in the latter case, is nothing but a specific precondition of inflammation [347]. As an allergen triggers inflammation in a sensibilized organism, there occurs a more or less expressed disorder of microcirculation in lungs with exudation and edema development, leukocytes migration to the area of inflammation, and proliferation of local cellular elements. Metabolites of AA, which are a central mediator in inflammation reaction, play an important role in the occurrence and supporting of these processes (Figure 2.4). The action of eicosanoids in inflammation is first of all directed at leukocytes. PGD_2, TxA_2, and especially LT B_4, 5-HETE, 15-HETE are potent chemoattractants and therefore play an important role in mechanisms of leukocyte infiltration. Vascular permeability is also increased by both PGs (E_2, D_2 and I_2) and LTs (C_4, D_4 and E_4), which generate the slow-reacting substance of anaphylaxis (SRS-A) as well as mono hydroxides of eicosanoids (5-HETE and 15-HETE). PGs act as potent vasodilators, whereas LTs increase vascular permeability by means of direct constriction of endothelial cells [581]. A damage of vascular permeability in combination with platelets aggregation caused by Tx, and a spasm of pulmonary arterioles under the impact of TxA_2 and PGD_2, may add to microcirculation dysfunction in lungs and the development of pulmonary hypertension.

Many pulmonary diseases are accompanied by the obstruction of the bronchial tree, pathogenesis of which is markedly affected by eicosanoids. Thus, a bronchospasm may be caused by PGD_2, $PGF_{2\alpha}$, TxA_2, LTB_4 and SRS-A, edema of bronchial mucosa may be caused by LTB_4, SRS-A, PGE_2, PGD_2, and mucous hypersecretion is increased by SRS-A, HETE. Besides taking part in these processes directly, eicosanoids either potentiate or inhibit effects of other biologically-active substances: histamine, bradykinin, serotonin, and acetylcholine. It is known that PGD_2 increases the release of histamine by basophiles, PGE_1 selectively potentiates the effect of bradykinin on vascular permeability, and $PGF_{2\alpha}$ inhibits an increase of vascular permeability caused by serotonin [646]. PGs play a key role in modulating the inflammatory process by providing two-way regulation of bronchial and vascular tone, migration, and degranulation of leukocytes and phagocytosis.

The involvement of AA metabolites in inflammation in acute and chronic diseases of lungs and bronchial epithelium has been demonstrated by many researchers. According to a hypothetic model [215], there is a kind of "nervous chain" in which the cells of the bronchial epithelium act as primary effectors that activate the LO pathway of AA metabolism in response to the action of pathological agent (e.g., respiratory virus)

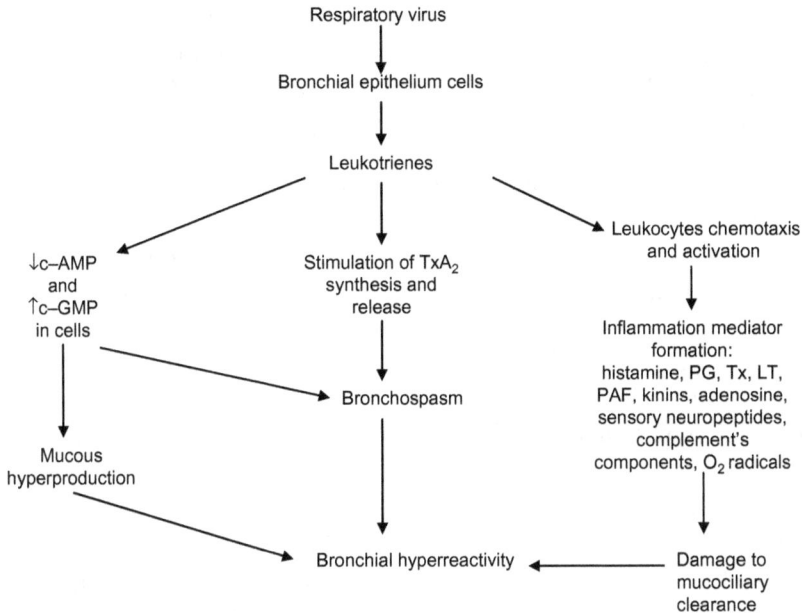

Figure 2.5 Hypothetical model of bronchial hyperreactivity formation in respiratory viral infection. From Gutkowski [215]

(Figure 2.5). LTs indirectly affect the obstruction of airways and the formation of bronchial hyperreactivity.

The role of AA metabolites has been more thoroughly studied with regard to occurrence and progression of bronchial asthma. Thus, it has been shown that in asthma the expression of both COX enzymes (COX-1 and COX-2) is increased in the bronchial epithelium and sputum cells. The expression of COX-2 increases in bronchial smooth muscle cells under the influence of proinflammatory ILs (IL-1β), which leads to production of large quantities of PGE_2. The latter has anti-inflammatory and bronchoprotective effect in the airway. Thus, in clinical investigations, inhalation of PGE_2 prevented bronchoconstriction caused by an allergen as well as an immediate and delayed asthmatic reaction, which was followed by a decrease of eosinophils in sputum [560]. PGE_2 also has proved to be effective in preventing a bronchospasm caused by physical exercise, although it did not affect the hyperreactivity of the bronchial tree in response to melacholine.

Mast cells are a main source of PGE_2 in lungs; they release it upon activation during both IgE-dependent and IgE-independent reactions.

PGD_2 is known to support inflammation in airways (Table 2.3). It is interesting that inhalation of PGE_2 has decreased an allergen-induced release of PGD_2 in airways in patients with asthma [564], which proves the existence of mutual control among AA metabolites.

Allergic inflammation in bronchial asthma is accompanied by excretion of both COX and LO metabolites of AA (LTs and LXs). LTs participate in main processes that exist in the human organism during an asthmatic reaction, including a spasm of bronchial smooth muscles (PG and TxA_2, SRS-A), mucosal edema in the bronchial tree (LTC_4, LTD_4 and LTE_4, PGE), increased mucosal secretion (PG, HETE), and cell infiltration of the airway wall with development of bronchial hyperreactivity (LTB_4, HETE) [65,167]. At the same time, LXs take part in resolution of inflammatory process [391,519,612]. They regulate chemotaxis, adhesion, and migration of leukocytes; block the leukocyte response to LTs (LTB_4, LTD_4); and support the phagocytosis of apoptotic leukocytes by macrophages [84,134,519]. As we know, a chronic allergic inflammation in asthmatics may result in structural changes in the bronchial wall, such as its thickening, subepithelial fibrosis and hyperplasia of mucous glands, myofibroblasts, bronchial smooth muscles, and vessel wall. A substantial role in this process of airway remodeling is played by sulfidopeptide LTs [225]. It has been demonstrated that in patients with severe asthma, along with the increased production of cysteinyl LTs, there is a substantial decrease in LX A_4 generation and its receptor gene expression in peripheral granulocytes, which is an evidence of a defect in negative control over the formation of inflammation mediators [440,441].

Eicosanoids production in different cells depends not only on the concentration of free AA, the activity of enzymes involved in their synthesis, but also on the effect of other signaling molecules and hormones of DNIES on this process. The data which have been obtained in recent years point to an important role of MT in that process, which affects AA metabolism at almost all stages. It has been established that MT inhibits the release of AA, which is followed by a decrease in the level of cytosolic phospholipase A_2 and mRNA expression [307]. In a similar manner, MT affects the LO pathway of AA metabolism. By affecting the high-affinity nuclear receptors, the hormone causes tonic inhibition of 5-LO gene expression both in the CNS [544] and on the periphery in B lymphocytes [659], decreases the expression of

12-LO, and thereby reduces the formation of 12-HETE [648,649]. MT has an opposite effect on the COX pathway of AA metabolism. Experimental studies have proved that it enhances the synthesis of PGE_1 and normalizes PGs production [234]. At the same time, MT's metabolite N^1-acetyl-5-methoxykynuramine reduces the activity of COX-2 [119,222].

MT not only changes the activity of main enzymes in the eicosanoids synthesis but also the nature of their effect. It reduces the vascular permeability caused by LTB_4, inhibiting the hyperadhesiveness of endothelial cells induced by this LT [315]. The AA metabolites, in their turn, are able to change MT production. Experiments on rat pineals have demonstrated that the intermediate metabolites of AA 12-HPETE and 15-HPETE increase MT production by stimulating the activity of NAT, the main enzyme of its synthesis [493]. Therefore, taking part in AA metabolism, MT has a regulating effect at the cellular level by regulating intercellular interaction and homeostasis. As we have already noted, the existence of pathology of all functional systems in patients with AIA points to the involvement of DNIES in mechanisms of this disease. Because of the broad action of MT on integrating all biological functions in this system, it is necessary to study this hormone production during AIA, even more so because one MT metabolite is a substance that is chemically similar to, and has the same effects as, ASA. Therefore, in Chapter 3, we describe the results of our own investigations and data contained in literature concerning the DNIES state in this disease.

State of DNIES in Aspirin-Induced Asthma

3.1 MT PRODUCTION IN PATIENTS WITH AIA

No data on MT production in patients with AIA have been reported in the available literature. Only one instance is mentioned about the decrease of MT levels in patients with bronchial asthma in exacerbation of the disease [212].

In the period from November 1, 1996, through to February 15, 1997, we studied the urinary excretion of 6-sulfatoxymelatonin (α-MT6s), the main metabolite of MT. As we know, the circadian excretion of α-MT6s adequately reflects MT production in the human body [61,62]. A total of 80.7% of this metabolite is excreted with urine in the period from 10 p.m. to 10 a.m., reflecting the circadian rhythm of MT production.

Given this, we studied the MT production by determining the urinary excretion of α-MT6s in daytime (from 9 a.m. to 9 p.m.) and nighttime (from 9 p.m. to 9 a.m.) by enzyme immunoassay with commercial kits of the reagents (DRG Instruments, Marburg, Germany), according to the methodology described in the documentation provided. For the α-MT6s measurements, the urine was collected into plastic containers. The volume of each sample was recorded, and 5-mL aliquots were stored at $-20°C$ until the analysis.

In the period of examination, the patients with bronchial asthma received regular asthma therapy (sodium cromoglycate, inhaled β-adrenergic agents, oral theophylline, and inhaled corticosteroids with daily dose of budesonide ≤ 800 µg). For a week before the examination, neither patients nor healthy subjects received any medication that could affect MT production (benzodiazepines, NSAIDs, dopamine precursors, and β-adrenoblockers), and kept a rest regimen (going to bed before 11 p.m.).

When selecting patients with bronchial asthma and healthy subjects, we tried to take into account not only the severity of the disease but also their age, as MT production in humans does not depend on gender but tends to decrease at the age over 45 years [21,474].

Urinary excretion of α-MT6s was studied in 13 patients with AIA (11 women and 2 men), of the average age 46.2 ± 2.6 years, of which 54% were under 45 years. A mild persistent clinical course of the disease was observed in three patients, the rest had moderate persistent clinical course of the disease. The group of patients with ATA also consisted of 13 subjects (5 women and 8 men), of the average age 40.2 ± 4.4 years, of which 61.5% were below 45 years. A mild persistent clinical course of the disease was observed in two patients, the rest had moderate persistent clinical course of the disease. The control group consisted of 12 healthy subjects (10 women and 2 men), of the average age of 41.5 ± 3.5 years, of which 58% were under 45 years.

Thus, the groups were practically homogeneous with regard to the age and disease severity.

The study has demonstrated that AIA patients differ from the ATA patients and control subjects in the lower daytime level of urine α-MT6s excretion (Table 3.1). As evident from Table 3.1, due to this fact, the patients with AIA demonstrate more expressed differences between diurnal and nocturnal urinary excretion of α-MT6s, while in the patients with ATA there is no such difference because of a high level of diurnal α-MT6s excretion. In 6 of 13 patients with ATA, the MT production at daytime was $2-2.5$ times higher than that at nighttime. There was no correlation between α-MT6s excretion and the age of asthmatic patients, and in healthy subjects it was weak ($R = -0.4$), which is probably due to the homogeneous age group of examined subjects.

Table 3.1 Mean $\pm$ SEM Level of Urine α-MT6s Excretion (ng mL^{-1}) in Asthmatic Patients and Control Subjects

Time	Groups			P_1	P_2	P_3
	AIA	ATA	Control Subjects			
9 a.m.–9 p.m.	12.7 ± 2.7	32.7 ± 9.9	19.9 ± 5.8	<0.05	<0.05	>0.1
9 p.m.–9 a.m.	33.7 ± 6.8	32.4 ± 7.4	45.1 ± 10.1	>0.2	>0.2	>0.2
P_4	<0.01	>0.1	<0.01			

Note: P, *significance of differences from Kolmogorov–Smirnov two-sample test.*
P_1 *for difference between AIA patients and control subjects,* P_2 *for difference between AIA and ATA patients,* P_3 *for difference between ATA patients and control subjects, and* P_4 *for difference between the indices of diurnal and nocturnal;* α-*MT6s excretion with urine.*

Given the fact that the debut of AIA usually occurs in humans younger than 45 years, we have separately analyzed the MT production in this age group of asthmatics and healthy subjects. It has been found that α-MT6s excretion in AIA patients of that age already at daytime was as low as 10.4 ± 1.6 ng mL^{-1}, while in ATA patients and in healthy subjects it was higher and amounted to 28.5 ± 14.4 and 23.8 ± 9.5 ng mL^{-1}, respectively ($P < 0.05$).

Thus, AIA patients showed decreased MT production at daytime, while ATA patients demonstrated a circadian rhythm disorder due to increased MT production at daytime.

It is known that the MT production from serotonin in the human body to a large extent (90%) depends on the activity of the enzyme NAT [239]. In order to find out whether this factor plays a role in changing MT synthesis in patients with bronchial asthma, we, in the period from November 1, 1997, through to February 15, 1998, investigated again the urinary excretion of α-MT6s at daytime and nighttime and the activity of NAT at daytime in AIA and ATA patients and healthy subjects of average ages 46.1 ± 2.6, 49.1 ± 4.5, and 40.6 ± 2.8 years, respectively ($P > 0.1$).

The group of AIA patients consisted of 18 subjects, of whom one had a mild persistent course of the disease, in 14 patients the disease was of moderate persistent severity, and 3 patients had a severe course of the disease. The number of ATA patients was 12, of whom 4 had a mild course of the disease, and in the others the disease severity was moderate. The control group consisted of 10 healthy volunteers.

AIA and ATA patients received a regular asthma therapy. Only two AIA patients received 10 mg of oral prednisolon per day.

The study has shown that the NAT activity rates in AIA patients is not different from that in ATA patients and healthy subjects, amounting to $14.77 \pm 1.95\%$, $13.88 \pm 2.29\%$, and $13.08 \pm 2.58\%$ ($P > 0.1$), respectively. It should be noted that the patients with bronchial asthma, as well as healthy subjects, had a slow type of acetylation (Table 3.2). The enzyme activity also was low (from 10% to 19%) in two patients who received oral GCs.

The comparison of enzyme activity with the age has shown that in healthy humans there was a direct correlation ($R = 0.6$, $P < 0.05$), and no such correlation was observed in patients with bronchial asthma.

Table 3.2 Distribution of AIA Patients, ATA Patients, and Healthy Subjects Depending on NAT Activity in Blood Based on Age

Group	NAT Activity (%)					
	0−9	10−19	20−29	30−39	40−49	50−59
AIA $N = 18$	4(22.2)	10(55.6)	3(16.7)	1(5.5)	0	0
ATA $N = 12$	4(33.3)	5(41.7)	3(25.0)	0	0	0
Healthy $N = 10$	3(30.0)	6(60.0)	0	1(10.0)	0	0

Furthermore, it turned out that diurnal urinary excretion of α-MT6s in healthy subjects correlates with the diurnal NAT activity ($R = 1.0$, $P < 0.05$). No such correlation was observed in AIA and ATA patients.

Thus, in the period of shorter daylight (November−February), there is a low activity of NAT in both asthmatic patients and healthy subjects. The change of MT production in AIA and ATA patients at daytime is not related to the enzyme activity but most probably depends on the state of extrapineal sources of MT.

MT synthesis in pinealocytes of the pineal gland is known to decrease abruptly at light, and probably the MT production at that time is determined by the state of MT-producing cell structures, with a leading role played by platelets which (like enterochromaffin cells in the gut) form a major peripheral reservoir of serotonin [368,580]. Platelets contain enzymes which are necessary for serotonin and MT synthesis [295]. As the uptake of serotonin by platelets is substantially reduced in patients with AIA [329], we may assume that this leads to a decrease of MT synthesis in them and to a large extent determines the low diurnal urinary excretion of α-MT6s.

We measured MT expression in platelets by indirect immunofluorescence assay [151,162] and determined that in healthy subjects $97.7 \pm 0.6\%$ of platelets showed MT-specific luminescence with intensity of 2.15 ± 0.01 A.U. At the same time, in AIA patients only $13.0 \pm 1.3\%$ of platelets had MT-specific luminescence ($P < 0.001$), with intensity of 2.14 ± 0.02 A.U. There was no MT expression in the rest of platelets of patients with AIA (Figure 3.1).

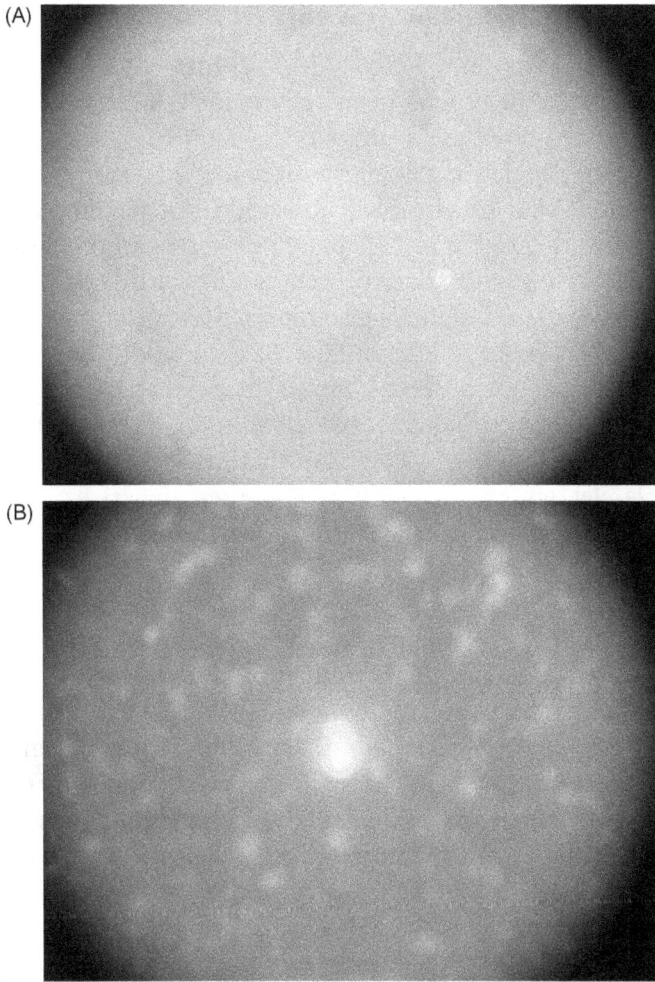

Figure 3.1 Immunohistochemistry showing the absence of MT expression in platelets from AIA patient (A) in comparison with the great number of luminous platelets from healthy subject (B).

Thus, patients with AIA characteristically have low content of extrapineal MT due to a decrease in its production by platelets, which in healthy humans not only generate up to 40% of extrapineal MT but also deliver it wherever and whenever necessary. Because MT is a regulator of platelet-vessel hemostasis, it was necessary to investigate peculiarities of platelets functional activity in patient with bronchial asthma and to find out whether it is affected by low MT production in patients with AIA.

3.2 PLATELET-VESSEL HEMOSTASIS IN PATIENTS WITH AIA

3.2.1 Platelet Functional Activity in Patients with AIA

Investigations of platelet functional activity in AIA began more than 30 years ago. Given the disaggregating effect of ASA, researchers studied its influence on platelets in patients with increased aspirin sensitivity. In 1974 Fisherman and Cohen [183] observed a longer time of capillary bleeding from the earlobes in AIA patients after ingestion of small doses of aspirin. However, this has not been confirmed by other researchers [557]. In 1979 Yalkut et al. [635] also reported a shorter time of capillary bleeding in AIA patients after intake of a threshold dose of aspirin, although they did not observe changes in number of platelets in the blood. The authors concluded that small doses of ASA have no effect on the platelet function. In the literature there is little and contradictory data on platelet aggregation activity. Some researchers have pointed to absence of changes in the intensity of platelet aggregation induced by ADP, adrenalin, AA in AIA patients [32,403]. At the same time, the PAF-induced platelet aggregation is credibly higher in patients with AIA than in healthy subjects, but not different from that in patients with atopic bronchial asthma [140]. Investigations of NSAIDs effect on platelet aggregation induced by ADP, AA, and collagen have demonstrated a high sensitivity of platelets to inhibitory effect of these agents in AIA patients [34,403,514]. In those cases, platelet aggregation was completely suppressed, and large doses of the said inductors were required to restore it. Changes in platelet functional activity in AIA patients have also been confirmed by the fact that platelets proved to be the only cells which excreted cytotoxic mediators and oxygen radicals in such patients in contact with ASA and other NSAIDs *in vitro* [12,13,227]. However, a study of 12-HETE production in platelets before and after intake of threshold dose of ASA has not revealed any relevant differences between AIA patients and healthy subjects.

Taking into account the revealed disruption of MT production by platelets in patients with AIA, we have also investigated the platelet functional activity in such patients and the influence of MT and ASA on this activity. For that purpose, we have studied platelet aggregation induced by ADP and heparin in 35 patients with AIA, 17 aspirin-tolerant asthmatics, and 22 healthy subjects by using C.R. Born's photometric method.

The analysis of ADP-induced platelet aggregation has shown that the amplitude of the total aggregation was increased in all patients with bronchial asthma as compared to healthy subjects [165]. This is mainly due to a change of the first aggregation wave: its amplitude is increased in patients with bronchial asthma, the more so in AIA patients ($P < 0.05$). The individual analysis of aggregation curves has shown that high values of amplitude of the total aggregation were observed in almost 40% of AIA patients and much less frequently in ATA patients and healthy humans. The biggest differences between the groups of patients were revealed in the analysis of the rate of first aggregation wave: its maximum values were 3 times more frequently observed in AIA patients as compared to ATA patients ($P = 0.03$) and 3.8 times more frequently than in healthy subjects ($P < 0.01$). Thus, the aggregation intensity in AIA patients increased at the expense of its first phase. A similar tendency was revealed by the analysis of heparin-induced platelet aggregation.

We then compared the platelet aggregation activity and MT production in patients with AIA and ATA and in healthy subjects. The individual analysis of correlations between the indices of ADP-induced platelet aggregation, which characterize its intensity, rate, and duration, and α-MT6s excretion level showed that in healthy subjects higher was the nocturnal level of α-MT6s the shorter was duration of the first phase of aggregation process registered in the morning: $R = -0.7$ ($P = 0.02$). At the same time, there is no such correlation in AIA patients, but some other relationships can be revealed: the higher the nocturnal MT production, the higher the rate of the first phase of platelet aggregation: $R = 0.7$ ($P = 0.02$) and its second wave: $R = 0.6$ ($P = 0.05$). There were no correlations in ATA patients [168].

Thus, in the case of increasing nocturnal MT production, it has a modulating effect on the platelet functional activity. However, if it has an effect on processes related to mobilization of calcium ions from intracellular stores in healthy subjects, due to which the first phase of ADP-induced platelet aggregation becomes shorter, then in AIA patients, as MT production increases, the platelet aggregation is reinforced because of its higher rate, which is caused by the inductor binding to receptors on the platelet membrane. In ATA patients, an equally-high MT production was observed at daytime and nighttime, without any correlation between its level and platelet functional activity.

It is known that on the human platelet membrane there are highly sensitive acceptors to MT, the state of which can be substantially changed in case of pathology of the platelet membrane–receptor complex. Given the obtained data, we have attempted to find out if there are any peculiarities of platelet reaction to exogenous MT in AIA patients.

For that purpose, we have studied the effects of different doses of MT *in vitro* on ADP and heparin-induced platelet aggregation in 15 patients with AIA, 17 patients with ATA, and 16 healthy volunteers. We simultaneously determined their diurnal and nocturnal urinary excretion of α-MT6s. The groups were uniform with regard to their age range and disease severity. The investigation was carried out in the period from November 1, 1996, through to February 15, 1997.

Platelet-rich plasma (PRP) was preincubated by MT doses corresponding to its physiological concentrations in the human body (1.0 and 100 pg mL^{-1} plasma), the dose of MT that produces an effect at the subcellular level (0.01 and 0.1 pg mL^{-1} plasma), and the pharmacological MT concentration (10000 pg mL^{-1} plasma).

The study of ADP-induced platelet aggregation has demonstrated that the preincubation with MT at a concentration of 0.01 pg mL^{-1} plasma, that is, 100 times lower than physiological concentration, causes a change of platelet functional activity in all examined subjects, although the character of changes in patients with bronchial asthma is significantly different.

Thus, in healthy subjects, an extension of latent period from 1.67 ± 0.87 s to 3.56 ± 0.73 s ($P < 0.05$) and the first phase from 49.89 ± 7.13 s to 55.78 ± 7.71 s ($P < 0.05$) was observed, which points to retardation of the first phase of aggregation process. The aggregation intensity increased in ATA patients at the expense of its second wave, as evidenced by the increase of the amplitude of total aggregation from 33.21 ± 12.05 to 55.21 ± 12.50 mm ($P < 0.05$) and its second wave amplitude from 8.00 ± 4.35 to 24.93 ± 9.77 mm ($P < 0.05$). In AIA patients, unlike ATA patients and healthy subjects, the incubation of PRP with MT at a dose of 0.01 pg mL^{-1} leads to an increased intensity of the first phase of ADP-induced platelet aggregation, as evidenced by the increase of the amplitude of its first wave from 32.44 ± 6.8 to 35.22 ± 6.56 mm ($P < 0.05$).

Preincubation of PRP with MT at 0.1 pg mL^{-1} plasma in AIA patients, as well as at a dose of 0.01 pg mL^{-1} plasma, leads to activation of the first phase of ADP-induced platelet aggregation: the rate of the first phase increased from 1.00 ± 0.22 to 1.10 ± 0.23 mm s^{-1} ($P < 0.05$). In the case of preincubation of plasma with MT at doses of 1.0, 100, and 10000 pg mL^{-1} in AIA patients, the aggregation parameters did not change as compared to initial values.

MT at doses of 0.1 and 1.0 pg mL^{-1} also did not cause any changes in ATA patients and healthy humans. However, preincubation with MT at a maximum physiological dose (100 pg mL^{-1} plasma) in ATA patients and healthy subjects leads to retardation of the first phase of aggregation process. Thus, ATA patients showed a shortening of the time interval from the addition of the aggregation inductor to the maximum amplitude of the first aggregation wave from 44.88 ± 2.69 to 41.00 ± 2.88 s ($P < 0.05$), and in healthy subjects the duration of the first phase increased from 50.33 ± 5.96 to 65.00 ± 4.12 s ($P < 0.05$). As for the MT dose of 10000 pg mL^{-1} plasma, it reduced the platelet aggregation intensity and its first wave in healthy humans only: the maximum amplitude of the total aggregation decreased from 53.96 ± 8.80 to 41.61 ± 7.04 mm, and its first wave amplitude decreased from 36.39 ± 5.21 to 30.50 ± 3.89 mm ($P < 0.05$).

Thus, it is in AIA patients that the addition of MT *in vitro* increases the intensity and rate of the first phase of ADP-induced platelet aggregation, but only at minimum concentrations—100 and 10 times lower than physiological ones. It should be emphasized that the initial platelet aggregation induced by ADP in AIA patients differs from that in ATA patients and healthy subjects in a higher intensity at the expense of its first phase.

Addition of MT *in vitro* in ATA patients at concentration which is 100 times lower than physiological one increases its intensity but at the expense of the second phase of aggregation, whereas when added at a maximum physiological concentration, it retards the first phase of ADP-induced platelet aggregation.

In healthy humans, MT retards ADP-induced platelet aggregation, and its inhibiting effect becomes more expressed with higher doses.

Furthermore, we have tried to find out whether the demonstrated platelet reaction to addition of MT *in vitro* correlates with the level of endogenous MT in patients and healthy subjects at daytime and at nighttime. As it turned out, the change of platelet reaction to the inductor after preincubation with different doses of MT in AIA patients was not related to MT production in humans. In healthy humans, however, after preincubation with MT at physiological concentration ($1.0 \, \text{pg mL}^{-1}$ plasma) or close to it ($0.1 \, \text{pg mL}^{-1}$ plasma), parameters of ADP-induced platelet aggregation correlated with endogenous MT production in daytime and at nighttime: the higher the level of α-MT6s in urine at daytime and nighttime, the lower the intensity of ADP-induced aggregation at the expense of the amplitude and rate of its second phase ($R = -0.8$, $P < 0.05$; $R = -0.9$, $P < 0.01$, respectively). In ATA patients, higher MT production at daytime correlates with a lower intensity of the first wave of ADP-induced platelet aggregation after preincubation only with physiological dose of MT— $1.0 \, \text{pg mL}^{-1}$ of plasma ($R = -0.8$, $P < 0.05$).

Consequently, only healthy humans and ATA patients demonstrate the inhibiting effect of endogenous MT on aggregation process which is observed after preincubation of PRP with physiological dose of MT. Absence of such effect in AIA patients is an indirect evidence of damaged reception of their platelets to endogenous MT.

To verify that the revealed peculiarities of platelet reaction to MT in AIA patients are regular and associated with their qualitative changes, we have studied the effect of MT $in \, vitro$ on heparin-induced platelet aggregation. For that we used the same doses of MT corresponding to its physiological concentration in the human body (1.0 and $100 \, \text{pg mL}^{-1}$ plasma), as well as a dose of $10000 \, \text{pg mL}^{-1}$ plasma which is 100 times higher than its physiological concentration.

It turned out that the preincubation of PRP with MT at a minimum physiological concentration ($1.0 \, \text{pg mL}^{-1}$ plasma) in AIA patients activates the first phase of aggregation; the time of maximum rate decreases from 255.00 ± 40.35 to 190.90 ± 25.69 s ($P < 0.05$) and the duration of the total aggregation shortens from 567.20 ± 65.97 to 472.10 ± 63.81 s ($P < 0.01$). In ATA patients, MT in the same dose inhibits the heparin-induced platelet aggregation, which is evidenced by a decrease of maximum amplitude of total aggregation from 73.80 ± 7.36 to 67.20 ± 7.90 mm ($P < 0.05$) at the expense of reduction of its second wave from 55.89 ± 8.75 to 47.78 ± 8.77 mm ($P < 0.05$).

In healthy subjects, no changes in aggregation process were observed in response to MT dose of 1.0 pg mL^{-1} of plasma.

The preincubation of PRP with MT at a maximum physiological concentration (a dose of 100 pg mL^{-1} of plasma) in AIA patients does not affect heparin-induced platelet aggregation, while in ATA patients the amplitude is observed to decrease from 55.89 ± 8.75 to 42.33 ± 10.07 mm ($P < 0.01$) and the rate of the second wave of aggregation decreased from 0.66 ± 0.17 to 0.51 ± 0.16 mm s^{-1} ($P < 0.01$). In healthy subjects, the amplitude of the second wave and aggregation duration also decreases from 19.88 ± 7.87 to 18.31 ± 7.95 mm ($P < 0.05$) and from 427.50 ± 41.78 to 377.50 ± 36.6 s ($P < 0.05$), respectively. An MT dose of 10000 pg mL^{-1} of plasma causes no changes in heparin-induced platelet aggregation either in asthmatics or in healthy humans.

Thus, it is only in AIA patients that the addition of MT *in vitro* increases the rate of the first phase of heparin-induced platelet aggregation, and only at minimum physiological concentration [176]. In ATA patients and healthy subjects, on the contrary, MT at physiological concentrations inhibits the second phase of platelet aggregation.

The revealed changes in platelet reaction to heparin after *in vitro* preincubation of PRP with MT at a dose of 1.0 pg mL^{-1} of plasma in patients with AIA do not correlate with the level of α-MT6s urinary excretion, that is, with endogenous MT production. The platelet aggregation parameters after preincubation with MT at a dose of 100 pg mL^{-1} of plasma correlated in ATA patients and in healthy subjects with nocturnal MT production, and in ATA patients they were reverse. In AIA patients, correlations are evident only when adding MT *in vitro* at a dose of 10000 pg mL^{-1} of plasma, namely, the decreased duration of platelet aggregation correlates with the decreased α-MT6s urinary excretion at daytime, and the reduction of the maximum rate of aggregation correlates with a decrease of nocturnal metabolite excretion. Consequently, in ATA patients and in healthy humans, it is the nocturnal MT production that can affect the functional state of platelets, while in AIA patients the diurnal MT production has a significant effect, and platelet aggregation is substantially activated in case of its decrease.

Thus, in AIA patients, a specific reaction to exogenous MT consists in the activation of the first phase of ADP and heparin-induced platelet aggregation, but only in response to addition of minimum doses of MT (0.01 and 0.1 pg mL^{-1} of plasma in case of ADP induction and 1.0 pg mL^{-1} of plasma in case of heparin-induced aggregation). One and the same regularity revealed by using two different inductors points to the defect of the membrane−receptor complex in AIA patients. The absence of correlations between aggregation parameters after preincubation with MT and preceding nocturnal MT production points to a weakening of its controlling action on platelet-vessel hemostasis.

It has been reported that one of the MT metabolites is a substance having a chemical structure and effects similar to that of ASA [256]. Since the platelet reactivity in AIA patients to inductors (ADP and heparin) changed substantially after preincubation of PRP with minimum doses of MT, it was necessary to investigate how platelets in these patients would react to the same inductors after preincubation with minimum doses of ASA to which the AIA patients had an increased sensitivity. Additionally, it was important to find out whether their platelet sensitivity to ASA correlated with endogenous MT production.

At the first stage of our investigation, we studied how ADP-induced platelet aggregation would change under the influence of different doses of ASA in patients with bronchial asthma and in healthy subjects, and then we compared the aggregation parameters with diurnal and nocturnal α-MT6s excretion individually for each subject. We used the concentrations (0.004 and 0.008 mg mL^{-1} of plasma) that correspond to 20 and 40 mg when recalculated for the respective blood volume, that is, they are doses to which the bronchial tree of AIA patients most frequently demonstrated sensitivity in per oral provocation test with aspirin.

The ASA effect on platelet aggregation was studied in the same patients who had been subjects for investigation of exogenous MT influence and α-MT6s urinary excretion at daytime and nighttime.

The study of ADP-induced platelet aggregation after preincubation of PRP with ASA at 0.004 mg mL^{-1} of plasma showed that only in AIA patients the maximum amplitude of total aggregation decreased from 50.06 ± 10.74 to 45.39 ± 9.86 mm ($P = 0.02$) and its first wave decreased from 36.44 ± 6.59 to 33.56 ± 6.31 mm ($P = 0.04$). Such changes in ADP-induced platelet aggregation in AIA patients were also observed after preincubation with ASA at 0.008 mg mL^{-1} of plasma. In that case, not only the intensity but also the duration of aggregation process decreased from 138.11 ± 22.24 to 108.33 ± 18.78 s ($P < 0.05$). At the same time, the same doses of ASA in ATA patients and healthy humans had no effect on ADP-induced platelet aggregation.

Thus, in AIA patients, in response to preincubation of PRP with ASA there were observed changes of the same aggregation parameters as in response to preincubation with MT: the first phase of aggregation was suppressed under the action of ASA and activated under the influence of small doses of MT.

Furthermore, we attempted to find out if there was correlation between the parameters of platelet aggregation after preincubation with ASA and MT production in humans. The results of individual analysis of correlations between the parameters of ADP-induced platelet aggregation after *in vitro* preincubation with ASA at dose of 0.004 and 0.008 mg mL^{-1} of plasma and the level of α-MT6s excretion in patients with bronchial asthma and in healthy subjects have shown that there is no correlation in AIA patients, and even the parameters of the first wave of aggregation do not correlate with the level of α-MT6s. However, there are correlations with the level of α-MT6s in spite of an absence of significant changes in aggregation parameters after preincubation with ASA in ATA patients and healthy subjects. They were more expressed in ATA patients and related to the intensity and rate of aggregation. Changes in the latter, along with increased diurnal and nocturnal excretion of α-MT6s, point to the inhibitory effect of endogenous MT the production of which in this group of patients is high not only at night but also at daytime. In healthy subjects, there is a correlation between the parameters of the second wave intensity and the total duration of aggregation and nocturnal excretion of α-MT6s.

After obtaining the results confirming that in AIA patients ASA, as well as MT, at small doses has a certain effect on the first phase of ADP-induced platelet aggregation and that the platelet reaction to an inductor does not depend on the level of endogenous MT production, we compared the parameters of ADP-induced platelet aggregation, which were simultaneously registered in each examined patient after preincubation with ASA, with those after preincubation with MT. In all AIA patients the values of almost all aggregation parameters had a high correlation, while in ATA patients only some of the parameters correlated in their values. In healthy subjects, there were regular correlations between the parameters that characterize the intensity and rate of aggregation.

The revealed, abrupt decrease in MT production by platelets in patients with AIA determined some peculiarities of their functional activity. MT has been shown to block Ca^{2+} influx through voltage-gated channels in mammalian cells regulating the enhancement of intracellular calcium via an MT receptor coupled to G_i- and G_o-proteins and thereby decrease the functional activity of platelets [69,202,605,649]. In platelets, there are acceptors with high affinity to MT [600]. Although the origin and properties of most ligands' acceptors have not been thoroughly studied so far, it has been demonstrated that these acceptors in their primary structure may be completely identical to receptors [142]. One may assume that platelet acceptors to MT possess the properties and structure of receptors to this hormone, and as we know MT receptors are found in various cells within different parts of the CNS. Since platelets, as well as other cells in the nervous system, have similar reception to neurotransmitters, one may assume that there are universal mechanisms of MT action on platelets and other cells. Normally, MT activates potassium channels coupled to G-protein, thereby inhibiting the excitation of neurons [377]. It has been demonstrated that MT also produces inhibiting effect on calmodulin-dependent phosphodiesterase [52,399]. Other researchers point to the ability of MT to suppress the Ca^{2+} influx into the cell, at the same time reducing the concentration of cAMP [594]. They also emphasize some peculiarities of the MT effect on the voltage-dependent calcium channels in various types of cells [371], as well as the ability of this hormone to enhance the ADP ribosylation of G-proteins by preventing the hydrolysis of bound GTP into GDP [72]. As we know, a change in the state of the cell receptor apparatus plays an important role in the genesis of some diseases. For instance, a substantial decrease of G_i-proteins has been found in the retina of rats with diabetes [257].

The specific features of *in vitro* platelet reaction to MT do point to some defect in the platelet receptor system (G_i- and G_o-proteins), which may damage their binding to MT. However, in platelets there is a system of G_s-proteins, α-subunits of which can activate certain channels for Ca^{2+} [202,281]. It is probably via this protein system that minimum doses of MT affect the platelets of patient with AIA. The existence of damaged platelet reception to MT is also evidenced by a positive correlation between the rate of the first and second phases of ADP-induced platelet aggregation and the nocturnal content of α-MT6s in urine, as well as by the absence of any correlations in case of further addition of MT *in vitro*. The opening of receptor-controlled calcium channels as well as a low basal level of MT at daytime decreases the stability of platelet membrane and increases platelet aggregation. This assumption is confirmed by the data obtained by researchers who have found a high influx of ionized calcium into platelets of AIA patients even under the action of low concentrations of PAF [140], and demonstrated a paradoxical reaction of platelets in AIA patients to the addition of adrenaline and ADP leading to a substantial increase of cAMP level [32,33].

Experimental studies have demonstrated that in AIA patients, in whom MT production in platelets is decreased, the specific reaction of platelets to exogenous MT and ASA consists of alteration of the first phase of aggregation process: it is activated under the influence of MT at very low concentrations but suppressed under the influence of ASA at minimum doses. It is known that the first phase of ADP-induced platelet aggregation reflects the processes of the aggregation inductor binding to receptors on the platelet membrane, opening of receptor-controlled channels for Ca^{2+}, and/or calcium exit from the intracellular stores. Therefore, the revealed alterations of platelet reaction to ADP and heparin after preincubation of PRP with MT or ASA point not only to pathology of the membrane–receptor platelet complex but also to a damage of calcium homeostasis of the cell. This specific state of platelets in AIA patients is evidenced by a lack of correlations between changes in aggregation parameters in response to *in vitro* addition of MT or ASA and the content of α-MT6s in diurnal and nocturnal urine samples. The independence of platelet reaction from endogenous MT, as well as high correlations between the parameters of ADP-induced aggregation after addition of ASA and after preincubation with MT, points to a lack of controlling effect of endogenous MT on their functional activity which is characterized by a high tension and low adaptive capabilities.

Thus, AIA patients, unlike ATA patients and healthy humans, have increased platelet sensitivity to ASA and a distorted reaction to MT, which points to qualitative changes in these blood cells.

The MT synthesis in pinealocytes is known to decrease abruptly in the light, and at daytime the MT production is determined by the state of cellular structures in the DNIES, particularly platelets which are a major reservoir of peripheral serotonin [289,295]. This is also confirmed by our data on a high MT expression in platelets of healthy subjects as well as by evidences that in platelets not only MT is formed but also one of its metabolites—5-SNAS, which, like α-MT6s—is excreted with urine, but in contrast to the latter, there is no circadian rhythm in the excretion of SNAS.

The membrane−receptor complex is a universal primary responsive system which, to a large extent, depends on microviscosity of the membrane lipids [598]. An increase in rigidity of the membrane phospholipid structures, which is due to insufficient functioning of the AOS and accumulation of LPO products, leads to a disturbance of surface receptor exposure, affecting the platelet reactivity to inducers of their aggregation [130,356]. Antioxidant properties of MT have been discovered recently, and it has been proved to have a direct effect on free radicals, acting as a "scavenger" and stimulating glutathione peroxidase. Increase of this enzyme activity directly correlates with concentration of MT in tissues [401,471]. Having the antioxidant properties, MT, on the one hand, decreases the ability of free radicals to damage DNA structure, and on the other hand, it substantially increases fluidity and reduces microviscosity of platelet membranes [114,126]. Low MT production in AIA patients may weaken the antioxidant defense, which results in damaging not only functions of platelet receptors but also the channels of ion permeability and lipid-dependent enzymes. This is further confirmed by data on lower glutathione peroxidase activity and increase of oxygen free radicals production in platelets of AIA patients [330,511].

Physiological concentrations of MT are known to affect calcium metabolism in cells by blocking L-type voltage-sensitive calcium channels, changing the level of Ca^{2+}/calmodulin in the cell and calcium content in intracellular stores [474]. Lower MT production in platelets of AIA patients, as well as in other cells of the DNIES, may lead to a disorder in cellular calcium homeostasis. A consequence of this is deep alterations in platelet form and ultrastructure found in AIA patients [9].

It should be noted that the revealed peculiarities of ADP and heparin-induced platelet aggregation in AIA patients are associated not only with a low level of basic MT production, but, at first, also with the pathology of the membrane—receptor complex of platelets themselves, which is a basis for developing a specific reaction of platelets to ASA in this disease. We have demonstrated that it is only in AIA patients that the intensity of the first phase of ADP-induced platelet aggregation decreases under the action of minimum ASA concentrations. According to literature data, ASA inhibits aggregation by lowering the second phase intensity due to the suppression of TxA_2 synthesis [85]. At the same time, ASA acetylates serine at the active center of COX [257]. It has also been proved that ASA is a strong acetylating agent which, when interacting with any proteins, acetylates the end amino groups of lysine. What is the basis for the altered reaction of platelets to ASA in AIA patients? One may assume that the acetylation of proteins of the initially damaged platelet membrane—receptor complex contributes to suppression of Ca^{2+} released from intracellular stores and thereby lowering the intensity of platelet aggregation and its first wave.

It should be emphasized that it is only in AIA patients that we have found a high correlation between changes in parameters of ADP-induced platelet aggregation after addition of ASA and after addition of MT, and the alteration of the first phase of aggregation *in vitro* after preincubation with ASA or MT did not correlate with the level of endogenous MT production. The obtained data show that at low MT production in AIA patients have no controlling action on the functional activity of platelets which is characterized by a high tension and low adaptive capabilities. It is known that functional activity of platelets to a certain extent depends on maturity of megakaryocytes in the bone marrow [248,538], and it is MT that, by penetrating through the megakaryocyte membrane and accumulating in the nucleus, affects some phases of translation and transduction in these cells [126]. Disruption of MT production in the bone marrow of AIA patients may be a reason of platelet immaturity and alteration of their functional activity.

3.2.2 Peculiarities of Nitric Oxide, Thromboxane and Prostacyclin Production in AIA

NO, as we know, performs an important function of regulating various intra- and intercellular processes [337,597]. There is a NO-dependent mechanism of activating guanylate cyclase, an enzyme which is a catalyzer of biosynthesis of cGMP, a potent regulator of cell functional activity, including the platelet aggregation [23]. By studying the platelet functional activity in AIA patients, we have found a higher intensity of ADP and heparin-induced aggregation at the expense of its first phase, which may point to a disorder in the cell calcium homeostasis. It has been proved that NO takes part in regulating intracellular concentration of Ca^{2+}, by coordinating various calcium-dependent metabolic processes, including those in platelets [481]. Furthermore, it has been shown that MT also participates in NO production by inhibiting both constitutive and inducible isoforms of NOS [195,446]. Taking into account the obtained results and literature data, we set out to discover the peculiarities of NO production in AIA patients and to compare it with the excretion of α-MT6s, a major MT metabolite, and with the platelet functional activity.

For that purpose, we observed patients with bronchial asthma in the remission of the disease (16 patients with AIA and 14 patients with ATA) and 14 healthy subjects, and studied their nitrate and nitrite urinary excretion at daytime and at night by using the method described by Madueno and Guerro [321], as well as α-MT6s, and on the following day—ADP-induced platelet aggregation. The groups were practically homogeneous with regard to the age and disease severity. The study was conducted in the period from November 1, 1996, through to February 15, 1997.

It has been found that patients with bronchial asthma as well as healthy subjects had a circadian rhythm of urinary excretion of nitrates and nitrites with prevailing concentrations at night (Table 3.3).

Along with the tendency to increase content of ions NO_2^- and NO_3^- in diurnal urine of AIA patients, there was an abrupt increase in the nocturnal samples of urine as compared to those in ATA patients and healthy subjects. On the other hand, the nocturnal urine of ATA patients contained less NO_2^- and NO_3^- as compared to their level in AIA patients, but their diurnal excretion of nitrate and nitrites was higher than that in healthy subjects. Thus, unlike healthy people and ATA patients, patients with AIA have an increased NO_2^- and NO_3^- excretion, which is especially high at night.

The comparison of NO_2^- and NO_3^- excretion and α-MT6s content for each examined subject in the same urine samples at daytime and at

Time	Groups			P_1	P_2	P_3
	AIA	ATA	Healthy			
9 a.m.−9 p.m.(n)	17.8 ± 4.0(15)	21.4 ± 4.0(14)	10.7 ± 1.5(14)	>0.05	>0.05	= 0.01
9 p.m.−9 a.m.(n)	88.1 ± 12.7(16)	59.9 ± 9.9(14)	65.9 ± 8.9(14)	= 0.01	<0.05	>0.05
P_4	<0.01	<0.01	<0.01			

Table 3.3 Total Urinary Excretion of Nitrates and Nitrites (nmol/mL) in Patients with AIA

Note: P_1 means the significance of differences between AIA patients and healthy subjects, P_2—between AIA and ATA patients, P_3—between ATA patients and healthy subjects, and P_4—between concentration in diurnal and nocturnal urine samples.

nighttime showed that in healthy subjects and in ATA patients the nocturnal level of NO production directly correlates with diurnal MT production ($R = 0.7$, $P = 0.03$ in healthy subjects and $R = 0.6$, $P = 0.04$ in ATA patients). No correlation between these indices was observed in AIA patients [146].

As we know, ions of NO_2^- are not only the products of NO metabolism but themselves also play an important role in regulating Ca^{2+}-mobilizing cell system. Furthermore, NO and products of its metabolism (NO_2) may induce the spatial protein distribution resulting in higher transition of proteins from a soluble to membrane-bound state, in which the rate of enzyme reactions grows up significantly, along with growing ATP synthesis [481]. Given these data, we have analyzed correlations between nitrate and nitrite ($\sum NO_2^-$ and NO_3^-) in urine and indices of ADP-induced platelet aggregation before and after *in vitro* preincubation with different doses of MT and ASA.

The ADP-induced platelet aggregation in AIA patients was initially different with regard to the first phase activation. It was a change in the parameters of the first phase of aggregation, pointing to a dysfunction of the cell receptor's complex and receptor-operated channels for calcium ions that was found throughout the analysis of platelet aggregation after preincubation with MT and ASA. As it turned out, in patients with AIA there were no correlations between the indices of ADP-induced platelet aggregation and diurnal and nocturnal NO production. After preincubation with MT at doses of 0.01 and 0.1 pg mL^{-1} of plasma, where the activation of the first phase of aggregation was most expressed, there were also no correlation between the aggregation parameters and nitrate and nitrite content in urine. When a dose of MT was increased to $1.0 \, pg \, mL^{-1}$ of plasma, there appeared a direct correlation between the duration of the first phase of aggregation and the content of nitrates and nitrites in nocturnal urine $(R = 1.0, \, P < 0.01)$. When the dose was $100 \, pg \, mL^{-1}$ of plasma, the correlation became reversed $(R = -0.6, \, P < 0.05)$, and in response to the dose of $10000 \, pg \, mL^{-1}$ of plasma no correlations were revealed again. It should be noted that after preincubation with ASA at a dose of $0.004 \, mg \, mL^{-1}$ of plasma, there were no correlations with nitrate and nitrite content in urine either. The same regularity of correlations as in case of preincubation with an increased dose of MT was observed after preincubation with ASA in a larger dose— $0.008 \, mg \, mL^{-1}$ of plasma: the more NO metabolites were found in nocturnal samples of urine, the higher was the rate of the first phase and the whole platelet aggregation. The correlations between the parameters describing the aggregation rate and the nocturnal NO production, which is much higher in AIA patients, confirm the role of the latter in changing the qualitative characteristics of platelets. These data once again confirm the fact that the perverse reaction of platelets in AIA patients to endogenous and exogenous MT, as well as to exogenous ASA, is due to qualitative changes in platelets themselves which lack the ability of coordinated functioning with biorhythms of NO and its metabolites production in the human body, which contribute to maintaining the normal functional state of the cell.

In ATA patients, however, correlations are high between the content of nitrates and nitrites in diurnal urine and the indices of ADP-induced platelets aggregation, and the nature of these correlations (maximum amplitude, amplitude of the first aggregation wave, and rate and amplitude of the second wave) points to an increase of the intensity and rate of the aggregation with an increase of diurnal production of NO. Such correlations remain the same after *in vitro* preincubation of platelets with MT and ASA. At that, the higher the diurnal NO production, the higher the intensity and rate of the first phase of platelet aggregation, and the shorter is its duration. It means that the decrease of platelet sensitivity to the inhibitory action of aspirin in ATA patients may be associated with high diurnal NO production.

In healthy subjects, there are also correlations between the parameters of ADP-induced platelet aggregation and the diurnal production of NO, but they are not of such expressed and rigid character as those in ATA patients and relate only to the indices that characterize the rate of aggregation. Nevertheless, the existence of such correlations points to the involvement of NO in regulation of platelet functional activity. In case of a higher diurnal NO production, the rate of aggregation remains high even under the influence of small doses of ASA.

Thus, in healthy subjects and ATA patients, the platelet reaction to *in vitro* addition of ASA changes depending on the NO diurnal production; that is, NO influences the functional state of platelets. This influence is considerable in ATA patients in whom correlations are highly determinate and rigid, which points to a high degree of tension in the homeostasis regulation mechanism. At the same time, there is no such correlation in AIA patients in case of excessive NO production at daytime and especially at night, which indicates to qualitative disorders in the state of their platelets. Due to a steep decrease of MT production in platelets, it probably does not produce an inhibitory effect on NOS activity and NO production, while an excessive production of the latter in patients with AIA leads to disorders in the state of the platelet-vessel hemostasis, which results in a significant disorder of pulmonary microcirculation, as demonstrated by our investigations.

NO production is known to be closely related to the formation of PGs, of which the most important is prostacyclin [360,554]. Prostacyclin precludes platelet aggregation, prevents their adhesion to the vascular wall, and causes vasodilatation [237]. An opposite effect is produced by TxA_2, which forms during the platelet activation [23]. Therefore, it seemed important to determine the content of prostacyclin and Tx in blood plasma of patients with bronchial asthma.

In the literature, we have come across only one mention of a high basal level of Tx in PRP in AIA patients [35]. We determined the content of prostacyclin and Tx in blood plasma by using radioimmune method. Our findings have demonstrated significant individual variations of prostacyclin and Tx values not only in patients with bronchial asthma but also in healthy subjects. Therefore, we analyzed individually the number of platelets and the absolute concentration of prostacyclin and Tx in plasma of asthmatics and healthy subjects. It turned out that 31.1% of AIA patients had a higher content of platelets in blood ($> 300 \times 10^9 \, L^{-1}$), whereas among ATA patients the percentage was 17.4%, and only 9.5% among healthy subjects ($P < 0.03$ between the indices in healthy subjects and AIA patients). In most patients with bronchial asthma, we observed low Tx production (<400 pg mL^{-1}): 62.1% in AIA patients, 65.2% in ATA patients, and only 33.3% in healthy subjects ($P < 0.03$). At the same time, there was a low prosta-cyclin production (<400 pg mL^{-1}) in most AIA patients—83.3% as compared to 50% in ATA patients ($P < 0.04$) and 62% in healthy sub-jects. As most AIA patients have an increased number of platelets, then in most of them (88.9%) there is less than 2 pg mL^{-1} of Tx if cal-culated as per one platelet, while the same quantity of Tx per one platelet is found in 60% of ATA patients ($P < 0.05$) and 37% of healthy subjects ($P < 0.001$). As for PG I$_2$ production per one platelet, it does not exceed 2.0 pg mL^{-1} in 83.3% of AIA patients, which is sig-nificantly ($P < 0.05$) observed more frequently, than in ATA patients and healthy subjects.

Thus, the low Tx and prostacyclin production in AIA patients also points to pathology of the platelet-vessel hemostasis, which may greatly contribute to the development of significant disorders of the capillary circulation in the lungs and, as a result, expressed disorders of the respiratory function.

3.3 ADRENAL GLUCOCORTICOID's FUNCTION IN PATIENTS WITH AIA

It is known that the hypothalamic–pituitary–adrenal (HPA) axis, being a major structural and functional component of the DNIES, pro-vides coordination of general homeostasis mechanisms. Dysfunction of the HPA plays a substantial role in pathogenesis of bronchial asthma but the mechanisms of GC deficiency formation in that disease have not been properly studied [587,588,591]. In recent years, researchers have demonstrated an important role of MT in regulating various elements of the neuroendocrine system in health and disease [473,486]. Taking into account the obtained data that point to the formation of GC's dependence in patients with AIA at early stages of the disease, we have studied cortisol content in blood serum in 16 AIA patients and 19 ATA patients before and after carrying out dexamethasone tests in compari-son with urinal excretion of α-MT6s.

In the group of AIA patients, 5 were with mild course of the dis-ease, 10 with moderate course, and 1 with severe course. In the group of ATA patients, 7 were with mild course and 12 with moderate course. In most patients with bronchial asthma, we observed a mixed type of the disease with exogenic and endogenic variants. The patients were examined in the remission of bronchial asthma with continuous antasthmatic therapy excluding systemic GCs.

The examination showed that in AIA patients, the morning content of cortisol was not notably different from that in ATA patients, amounting to 444.9 ± 53.3 and 487.5 ± 78.6 nM L^{-1}, respectively. Individual analysis showed that in every fifth patient with bronchial asthma this content exceeded the maximum level in healthy subjects (600 nM L^{-1}).

After ingestion of 1.0 mg dexamethasone all patients with bronchial asthma, except one ATA patient, showed significant decrease of cortisol content in blood as compared to the initial value. Although its average values in the groups of examined patients were not reliably different because of high individual variations, there was an obvious tendency to lower cortisol content in AIA patients— 39.4 ± 5.4 nM L^{-1}, whereas in ATA patients it amounted to 60.5 ± 20.2 nM L^{-1}. Individual analysis showed that the decrease of cortisol content in blood exceeded 90% of the initial level ($P < 0.04$) in 72% of AIA patients and only in 21% of ATA patients.

Analysis of correlation between the cortisol content in blood before and after ingestion of dexamethasone showed that there was no such correlation in AIA patients, whereas it was evident in ATA patients ($R = 0.9$, $P = 0.004$), that is, we observed decrease in the regulatory control of cortisol content.

Diurnal urinary excretion of α-MT6s was 12.7 ± 2.7 ng mL^{-1} in AIA patients and 32.7 ± 9.9 ng mL^{-1} in ATA patients ($P < 0.05$), while nocturnal excretion was 33.7 ± 6.8 and 32.4 ± 7.4 ng mL^{-1} ($P > 0.2$), respectively. No correlations were found in all the patients with bronchial asthma between α-MT6s content in diurnal and nocturnal urine samples and cortisol content before and after ingestion of dexamethasone.

However, the comparison of each patient's reaction to dexamethasone with the content of α-MT6s in diurnal and nocturnal urinary excretion showed that the reaction to dexamethasone test was higher than 90% of the initial value in those patients whose diurnal urinary excretion of α-MT6s was lower than 16 ng mL^{-1}. As the content of α-MT6s in diurnal urine was low in patients with AIA that determined the characteristic decrease of cortisol content after ingestion of dexamethasone due to the loss of modulating role of MT on regulation of HPA axis.

The investigation of the adrenal GC function has proved that in AIA patients with low diurnal MT production in response to dexamethasone ingestion, the content of cortisol in blood decreases abruptly, and the decrease degree does not depend on its initial level. This may point to a disorder in pineal control of regulating the central parts of the HPA axis, which, under conditions of chronic stress in patients with AIA may contribute to the development of secondary insufficiency of the adrenal cortex GC function.

3.4 PECULIARITIES OF IMMUNE SYSTEM IN PATIENTS WITH AIA

Within the DNIES structure, the immune system plays a special role in supporting homeostasis, as its function is to identify extraneous molecules and destroy them. The emergence of AIA has been always viewed by clinicians in connection with intake of ASA, as clinical manifestations of ASA intolerance in aspirin-sensitive patients with bronchial asthma resemble immediate hypersensitivity reaction. Thus, within an hour after aspirin intake, patients begin to have acute asthma attack followed by rhinorrhea, lacrimation, and reddening of the upper body, mostly head and neck [573]. All these reactions are very dangerous, as they develop like lightning and can lead to faintness, shock, or even fatality [567].

In the past, there were numerous attempts to prove the allergic nature of AIA. Some researchers have obtained data confirming the involvement of IgE in the formation of hypersensitivity to NSAIDs [650]. Thus, high concentrations of common IgE and specific IgE antibodies to aspirin have been found in some patients with intolerance of ASA and other NSAIDs [40,187]. However, skin tests with aspirin proved to be negative in AIA patients, and the existence of hypersensitivity not only to ASA but also to other NSAIDs having different chemical structure discredits the allergic nature of the disease. At the same time, investigations of immune status in patients with asthmatic triad (aspirin-exacerbated respiratory disease, AERD) have demonstrated an increase in the level of proinflammatory cytokines synthesized by epithelial cells and activated by Th_2 lymphocytes (IL-2,

IL-3, IL-4, IL-5, IL-13, granulocyte-macrophage colony-stimulating factor (GM-CSF), and eotaxin (EX)) [38,39,220,540,541], as well as a tendency to lowering the level of T-lymphocytes due to a decrease of CD4(+) cells [372].

The results of our analysis have also demonstrated dysimmunity in patients with AIA. Thus, 85.7% of AIA patients at the age from 16 to 29 showed a certain correlation between asthma exacerbations and an AVRI. Furthermore, the existence of associated chronic inflammatory diseases of the upper airways and connection of asthma attacks with AVRI or pneumonia in almost all AIA patients point to insufficient anti-infection resistance in these patients. Therefore, we attempted to demonstrate the existence of chronic persisting infection of airways in patients with AIA.

In recent years, researchers have paid a special attention to studying the role of *Chlamydia pneumoniae* in the development of bronchial asthma [10,101,203,574]. Epidemiologic studies have shown that over 60% of adults suffer from the respiratory infection caused by this germ, and in 90% it has asymptomatic or subclinical course [506]. Breakouts of the respiratory infection caused by *C. pneumoniae* are observed for 2–3 years in 2–10 year intervals, every time affecting not only a large number of adults but children as well, and bringing along the development of bronchospasm, bronchial hyperreactivity, and chronic obstructive bronchitis [217,218]. More and more data have been accumulated recently, which point to relationship between chronic *C. pneumoniae* infection and appearance of bronchial asthma or its recrudescence, as well as atherosclerosis, myocardial infarction, and ischemic stroke [58,100,216,549,550].

According to the latest classification, *C. pneumoniae* belongs to the genus *Chlamydophila* which is genetically and phenotypically different from the genus *Chlamydia*, but evolutionarily akin to it in the structure of various genes including the genes of ribosome-operon and outer membrane proteins [136]. *Chlamydia* are gram-negative microorganisms capable of intracellular reproduction only, like viruses. Under conditions of decreased immunologic resistance, a durable persistence of the germ is formed in epithelial cells, alveolar macrophages, and fibroblasts of infected mucous membranes. Being absorbed by peripheral monocytes, they spread in the organism and settle down in vascular endothelium and other tissues, which lead to development of some autoimmune diseases [63,66,198]. Chlamydial infection, especially in case of dysbacteriosis, acquires a chronic recurrent course, creating conditions for propagation and persistence of other opportunistic pathogenic microflora, which results in frequent occurrence of acute respiratory diseases. Consequently, some researchers put a special emphasis on the role of chronic chlamydial infection in development of infection-dependent severe bronchial asthma [57,216,218].

Taking into account the literature data and results of our own investigations, we decided to study the frequency of *C. pneumoniae* infection in patients with AIA in comparison with some indices of the immune system and clinical course of the disease. More than 44 patients with AIA were examined. The control groups consisted of 65 aspirin-tolerant subjects with bronchial asthma and 21 healthy subjects. The patients were examined in the remission of bronchial asthma with continuous antasthmatic therapy excluding systemic GCs.

The content of IgA and IgG antibodies to *C. pneumoniae* and *C. trachomatis* in blood was evaluated by the enzyme immunoassay with commercial kits of the reagents. Furthermore, we evaluated the content of IgA, IgM, IgG, CIC, complement titer (CH50), as well as phagocytic activity of neutrophils, spontaneous migration, and migration inhibition index of granulocytes and monocytes.

The investigations showed that in bronchial asthmatics as well as in healthy people antibodies to *C. pneumoniae* are revealed much more frequently than antibodies to *C. trachomatis* (Table 3.4). At that, IgA antibodies to *C. pneumoniae* are revealed more frequently. It appears that in patients with AIA, the IgA titer much more frequently is equal to or exceeds 1:32. It should be emphasized that this antibody titer prevails in AIA patients (37.5% against 12.5%, $P < 0.001$), while in ATA patients IgA antibodies are revealed in a lower titer. In healthy subjects, except one case, the prevailing titer was 1:8 to 1:16. As for IgG antibodies in titers $\geq 1:16 < 1:64$, pointing to an earlier infection, they equally often occurred both in bronchial asthmatics and healthy subjects. Age distribution of patients with bronchial asthma who have antibodies to *C. pneumoniae* is shown in Table 3.5.

Apparently, in patients with AIA, IgA antibodies are revealed mostly at the age under 50, whereas in aspirin-tolerant asthmatics above 50 years ($P < 0.001$). The same regularity exists for IgG

Table 3.4 Frequency of Revealing Different Antibodies Titer to *C. pneumoniae* in Patients with AIA

Groups	IgA		IgG	
	≥ 1:8 < 1:32	≥ 1:32	≥ 1:16 < 1:64	≥ 1:64
AIA patients	12.5%	37.5%	22.7%	2.3%
ATA patients	32.7%	23.1%	15.4%	10.8%
Healthy subjects	38.1%	4.8%	19.0%	–
P_1	<0.001	<0.001	>0.1	–
P_2	>0.1	<0.01	>0.1	–
P_3	<0.01	>0.1	>0.1	= 0.03

Note: *Significance of differences between the groups: AIA patients and healthy subjects—P_1, ATA patients and healthy subjects—P_2, AIA and ATA patients—P_3.*

Table 3.5 Age Distribution of Patients with AIA Having Different Titers of Antibodies to *C. pneumoniae*

Age	IgA (≥1:8)			IgG (≥1:16)		
	AIA(%)	ATA (%)	Healthy (%)	AIA (%)	ATA (%)	Healthy (%)
20−29	8.3	13.8	11.1	9.1	17.6	–
30−39	16.7	10.3	33.3	18.2	17.6	50.0
40−49	66.7	13.8	33.3	54.5	17.6	25.0
≥ 50	8.3	62.1	22.3	18.2	47.2	25.0

antibodies. Healthy subjects with antibodies equally often occurred in all age groups. The existence of antibodies not only to *C. pneumoniae* but also to *C. trachomatis* was noted in one patient with AIA and in four ATA patients.

The comparison of the obtained data with peculiarities of the disease clinical course has shown that most AIA patients had a severe course of disease.

It is known that a chronic chlamydial infection substantially changes the activity of macrophages, decreases the level of immunoglobulins, and increases the content of CIC in blood. We compared measurements of immunity in patients with bronchial asthma with IgA antibodies ≥1:8 to *C. pneumoniae* (1st subgroup) and without antibodies (2nd subgroup). The number of CIC exceeded 0.1 A.U. in 44% of patients with IgA and in only 18.2% ($P = 0.03$) of patients without antibodies. The spontaneous migration of granulocytes and monocytes as well as the index of granulocytes migration inhibition were within normal values, but the index of monocytes migration inhibition was lower than normal in 59.3% of the first subgroup patients ($<30\%$), whereas this index was present in only 20.0% of the second subgroup patients ($P < 0.01$).

Thus, the high frequency of finding IgA at the age below 50 years in AIA patients, as well as a higher titer of antibodies as compared to than in ATA patients, point to the existence of a chronic persisting chlamydial infection which forms due to insufficient functions of the immune system.

3.5 AA METABOLISM IN PATIENTS WITH AIA

Literature data confirm that patients with AIA have a disorder of AA metabolism, and that changes affect both the COX and LO pathways. It has been established that they have a decreased expression of mRNA COX-1 and COX-2 (particularly COX-1), in epithelial cells of the bronchial tree [437,438,540].

At the same time, no statistically-relevant differences have been revealed by comparing the epithelial expression of both enzymes in patients with AIA and in aspirin-tolerant asthmatics [106]. Studies of the kinetics of COX-2 expression in nasal mucosa and polyposis tissues of aspirin-sensitive asthmatics have not shown any dynamics either, although in aspirin-tolerant asthmatics the COX-2 expression in polyps substantially increased during 60 min at room temperature [451,484]. These data may indirectly point to a disorder in the regulation of the activity of both COXs in aspirin-intolerant asthmatics.

Quantitative studies of AA metabolites have also demonstrated certain peculiarities in patients with AIA, which have bearing on changes in both the COX and LO metabolites. Thus, a high initial level of urinary LTE_4 has been noted in aspirin-sensitive asthmatics as compared to that in ATA patients [19,284,353,354,392,397], the high LTE_4 level being found in urine but not in sputum [229]. At the same time, when comparing the LTE_4 urinary excretion in 10 AIA patients and 31 aspirin-tolerant patients with bronchial asthma, other researchers have found that it was the same in both groups. Furthermore, no correlation has been revealed between the LTE_4 content in urine and the forced expiratory volume in 1 s (FEV_1), which, in the author's opinion, points to absence of influence of this AA metabolite on the level of bronchial obstruction in asthma patients [535]. Many researchers, however, have observed a higher content of 15-HETE, LTC_4, and LTE_4 both in urine and nasal and bronchoalveolar lavage (BAL) of patients with AIA after oral, nasal, or bronchial provocation tests [276,278,279,284,354,359,533]. The increase in LT content after aspirin provocation in these investigations was followed by a decrease in the content of COX products, such as PGE_2, $PGF_{2\alpha}$, TxB_2, and PGD_2, although the content of $9\alpha11\beta PGF_2$ and 6-keto-$PGF_{1\alpha}$ only tended to decrease. Some researchers have found an increase in PGD_2 production, which goes up after aspirin challenge [59,229]. O'Sullivan et al. [391] demonstrated a substantial increase in urinary excretion of the PGD_2 metabolite-9α-$11\beta PGF_2$ in patients with AIA after intake of a large dose of ASA. There were no statistically significant changes in the content of 5-, 12-, and 15-HETE in BAL in AIA patients after inhalation of lysine-aspirin [532].

On the other hand, the formation of PGE_2, which promotes bronchodilatation, was significantly decreased after stimulation by anti-inflammatory factors [438], although the level of this PG does not change after aspirin challenge [345]. This confirms that in aspirin-sensitive asthmatics the aspirin intake leads to inhibition of COX-1, but the formation of PGE_2 may remain under effect of COX-2, although the expression of its mRNA in nasal polyps of aspirin-sensitive asthmatic is decreased [435]. The demonstration by many researchers of a large number of LTs in AIA patients before and after exposure to ASA or other NSAIDs [113,532,563,564] has stimulated a search for a pathology of LO which is involved in their synthesis. Cowburn et al. [106] have found a large number of cells expressing LTC_4 synthase in bioptates of bronchial mucosa by using rabbit-refined polyclonal antibodies to enzymes of lung tissue in aspirin-sensitive asthmatics. The number of these cells was 5 times higher than that in aspirin-tolerant asthmatics and 18 times higher than in healthy controls. The immunostaining of 5-LO, its activating protein, LTA_4-hydrolase, COX-1, and COX-2 has not shown substantial differences between the groups of examined subjects. The number of cells expressing LTC_4-synthase correlated with the initial level of LTs in BAL and the degree of bronchial responsiveness to aspirin inhalation in AIA patients. Expression of LTC_4-synthase has been mostly observed in eosinophils (71%), and to a less extent in mast cells (<20%) and macrophages ($<10\%$). Similar data have been obtained in respect to polyposis tissue in patients with asthmatic triad [3]. Based on the obtained data, the authors have assumed that aspirin-intolerant patients either have genetically conditioned hyperexpression of LTC_4-synthase in eosinophils or acquired under the influence of IL-5 which is a selective chemoattractant for eosinophils and increase the ability of the latter to synthesize sulfidopeptide LTs *in vitro* [64]. The high expression of LTC_4-synthase in AIA patients may be a cause of constantly high production of LTs in these patients [230,495]. Aspirin intake, in its turn, inhibits the synthesis of PGE_2 and leads to excessive formation of LTs due to reduced control of the latter over their synthesis [501,615]. Further investigations have shown that there is genetic polymorphism of the fifth nontranslated region of LTC_4-synthase which consists in transversion of nucleotide 444 from allele A to C [501]. In AIA patients, allele C_{444} is found in 60% more often than in ATA patients. In AIA patients who are carriers of this allele there is a disorder of LTC_4-synthase expression, and the aspirin provocation test leads to very high urinary excretion of LTE_4 [563]. At the same time, it has been noted that almost 30% of AIA patients have no changes in the gene structure of LTC_4-synthase, and 25% of healthy subjects are carriers of C_{444}-allele without any unfavorable consequences for their health. Therefore, the existence of C_{444}-allele is not the obligatory sign of AIA.

At the same time, the studies of LT receptors on different cells have shown that AIA patients have an increased number of CD45 + leukocytes in the bronchial mucosa, which express one type of receptors to sulfidopeptide LTs (cysLT1) [104,539].

Thus, patients with AIA show disorders in the LO pathway of AA metabolism directed to increasing the LTs' production and reception to them.

In recent years, researchers' attention has been drawn to mechanisms controlling the increased formation of LO metabolites of AA and airway inflammation. It is known that PGE_2 limits LT synthesis in certain human cells, including eosinophils, neutrophils, and macrophages, inhibiting 5-LO [14,93,283]. It has been found that inhalation of PGE_2 blocks the increase of sulfidopeptide LTs synthesis induced by NSAIDs in AIA patients [81]. Furthermore, it has been shown that PGE_2 synthesis decreases in blood macrophages of some patients with

AIA in absence of aspirin ingestion [507]. A low concentration of PGE_2 also was observed in nasal polyps of these patients [642].

As we know, there are four receptors for PGE_2. It has been found that patients with asthmatic triad have a substantially lower number of neutrophils, mast cells, eosinophils, and T-cells which express the second type of PGE_2 receptor [641], and its transcription is reduced [244].

Thus, a disorder of PGE_2 synthesis and reception in patients with AIA can be viewed as a defect of negative control over formation of inflammation mediators in that group of patients.

The studies of other anti-inflammatory mediators, LXs, also have revealed their decrease in AIA patients in 2 h after aspirin provocation, in contrast to ATA patients in whom the intranasal administration of aspirin resulted in an increase of LX A_4 level [278,285,422,500,636]. Recently, yet another mechanism of ASA action has been found. The interaction between the aspirin-acetylated COX-2 in endothelial cells and leukocyte 5-LO leads to generation of 15-epilipoxins which, like LX A_4, inhibit neutrophil's adhesion and cellular proliferation [575], reduce the peroxynitrite formation in neutrophils, monocytes, and lymphocytes, and control many cellular functions of the immune system [22,249]. It is worth noting that LX A_4, as well as the ASA-triggered 15-epi-LXA$_4$, inhibits 5-LO and PGE_2 formation [441]. A decrease in LX formation after aspirin ingestion in aspirin-sensitive asthmatics points to a perverse reaction of cells to NSAIDs [499].

Given these data, Szczeklik et al. [559,564,566,571] have supposed that the formation of sulfidopeptide LTs, which have a potent inflammatory effect, is enhanced in patients with AIA due to reduced production of PGE_2 and LXs, which are functional antagonists of LTs, or due to abnormality of receptor sensitivity to these substances. Such disorders, as the authors believe, might have caused a past viral infection or unfavorable environmental conditions [563].

Thus, in spite of inconsistency of some data, numerous studies of AA metabolism in aspirin-sensitive asthmatics have demonstrated its regulatory disorder, enhancement of LO pathway, selective decrease in production of certain COX metabolites, and changes in the reception to PGs and LTs.

CHAPTER 4

Pathogenesis of Aspirin-Induced Asthma

Nowadays, aspirin and other NSAIDs are widely used in clinical practice. At the same time, the occurrence of various adverse reactions to these preparations among general population is rather high [219,270,560]. This has become a basis for elaborating a classification of allergic and pseudoallergic reactions to NSAID intake [204,545]. The following clinical variants have been proposed:

1. NSAID-induced rhinitis and asthma,
2. NSAID-induced urticaria and angioneurotic edema,
3. Single-drug-induced urticaria and angioneurotic edema,
4. Multiple-drug-induced urticarias and angioneurotic edema,
5. Single-drug-induced anaphylaxis,
6. Single-drug-induced or NSAID-induced mixed reaction.

This classification does not, however, take into account various pathogenic mechanisms underlying one-type clinical manifestations, which is necessary for developing effective methods of prevention and treatment. It is especially true of AIA. Numerous attempts to prove the allergic nature of this disease, based on IgE participation in the formation of hypersensitivity to aspirin and other NSAIDs [40,191,650], have failed to explain the whole range of AIA clinical manifestations in adults and children.

Our long-time clinical observations and investigations have demonstrated that, on the one hand, there is bronchial asthma with drug allergy to aspirin intake and generation of aspirin-specific IgE antibodies. This is an atopic (exogenic) form of bronchial asthma with allergy to different NSAIDs. On the other hand, there is a specific nosological form of such disease called "AIA," which, firstly, has a unique clinical presentation that makes it different from bronchial asthma with various clinical and pathogenic variants; second, its development is often preceded by Quincke's edema; and third, it is accompanied by pathologies in the upper airways (nasal polyposis, vasomotor rhinitis, chronic maxillitis, ethmoiditis) and in other functional systems of the human body (nervous, endocrine, immune systems). Given all this, AIA should not be viewed as an allergic disease, as it has its own specific path of physiological mechanism of development.

No commonly acceptable theory of AIA pathogenesis has been proposed so far. Major works by Russian and foreign researchers are mostly dedicated to studying specific aspects of AA metabolism, excessive LT production, and increased airway responsiveness to them [24,92,95,96,106,283,568,573]. Some researchers lay emphasis on the effects of a persisting viral infection changing the functions of lymphocytes and other cells (eosinophils, mast cells, basophils) within the quick response system that provide the local regulation of airway conduction [558,565,567,570]. All current hypotheses explain only the development of aspirin-induced bronchospasm but do not allow the formation of a clear view of pathogenesis of aspirin-induced bronchial asthma taking into account all its clinical manifestations which connect not only with the respiratory system, but also with other functional systems of the human body.

Numerous genetic studies have been conducted in recent years to find out those genes responsible for certain inflammatory disorders in AIA. However, the polymorphism of some genes revealed in aspirin-intolerant asthmatics in one population [106,415,501] was not found in the other [89]. The revealed genetic disorders in AIA are associated with alterations in one of the links of the inflammatory process [4,255,263,264,265,402,500,524]. Besides, a single genetic defect or promoter gene has not been found so far, which is responsible for appearance of numerous disorders in AIA [89,277,546].

We have attempted to summarize here the results of our own investigations and literature data in order to explicate the mechanisms underlying clinical manifestations of AIA.

Most Russian and foreign researchers, as well as studies conducted by the AIANE which was created in 1993 and united eight European countries, present statistical data concerning symptoms of aspirin and other NSAIDs intolerance, allergologic and hereditary history and efficiency of applied therapy in aspirin-intolerant asthmatics. Some authors have pointed out that, apart from asthma attacks in response to ASA intake, 36% of patients have rhinorrhea and edema of nasal mucosa, 30% have urticaria and Quincke's edema, and 4% have anaphylactic shock [386]. According to the same authors, only 5.2% of patients had close relatives with NSAID intolerance which occurred mainly in the form of rhinitis, shortness of breath, and sometimes urticaria. It should be noted that almost all researchers call attention to severe course of AIA, even without intake of ASA or any other NSAIDs [13,459,512,570].

The results of our investigations, however, confirm that unlike ATA patients, patients with AIA suffer combined damage to all vital organs and systems already at an early stage prior to the occurrence of asthmatic syndrome. Thus, symptoms pointing to CNS dysfunction are revealed very early in almost all AIA patients. AIA patients show early indications of psychological disorders, manifesting themselves as alarm-depressive or asthenoneurotic syndromes at a young age in 78.6% of patients, and its occurrence remains high in all age periods, pointing to increased anxiety, unformed value system, and a low ability to make independent decisions [333,445]. In view of the evolution theory, such an organism has the least perfect mechanisms of environment adaptation [280]. Indeed, according to our data, in 18.3% of AIA patients, the debut of disease was connected with a psychoemotional trauma, which is in line with the materials recently published by the AIANE, pointing out that psychological stress was an initiating factor of debut and recurrent attacks in 240 of 365 patients with AIA [386]. Furthermore, the occurrence of NARES, considered by Moneret-Vautrin et al. [366] as a pre-existing disease of aspirin triad, is also connected with certain stress situations in case history (mourning, occupational conflicts, divorces, unemployment, etc.) in a half of aspirin-intolerant asthmatics.

It is known that changes not only on the part of the CNS, but also of the autonomic nervous system, play a major role in the pathogenesis of bronchoobstructive syndrome in all bronchial asthmatics [180,296]. It is believed that imbalance between activating and an inhibitory effect of different elements of autonomic nervous system plays a significant part in the genesis of bronchial hypersensitivity and hyperresponsiveness which are distinguishing features of bronchial asthma. It is a common view that in bronchial asthma, there is an increased effect of the parasympathetic acetylcholine regulator and a dysfunction of adrenergic component: increase of α-receptors sensitivity and decrease of β-receptors sensitivity, such alterations being secondary against somatic pathology [369,490,652]. Our investigations have shown that nearly all AIA patients demonstrate clear signs of altered bronchial reactivity, which is confirmed by a link between asthma attacks and inhaling cool air and irritant odors in 93.3%, abrupt change of weather in 55.1%, and physical exercise in 73.9% of patients with AIA. Some researchers draw attention to the connection between recurrent attacks of AIA and the factors that we have pointed to [336,386], but we have found out that these symptoms of the bronchial hyperresponsiveness are observed in AIA patients already at the preasthmatic stage, which may be an evidence of an early manifestation of disorders in autonomic nervous system regulation. Our conclusion is supported by the data obtained by Moneret-Vautrin et al. [366], who have found a dysfunction of the autonomic nervous system in patients with the NARES. The authors have demonstrated an increased sensitivity to adrenergic agents at the level of β_2-receptors in large vessels and β_1-receptors in the heart according to the isoprenaline test, and a reactivity of α_1-receptors in small skin vessels with endermic administration of papaverine hydrochloride which causes edema and vasodilatation in healthy subjects. A study of the activity of lymphocyte β-adrenergic receptors, however, has not revealed any significant difference with those in healthy subjects [630]. The existing disorder on central regulatory mechanisms may play a certain part in the formation of a primary dysfunction of the autonomic nervous system. This can also be confirmed by the data obtained by Lovitsky [316] who has found electroencephalographic changes in patients with asthmatic triad, pointing to a disorder in intracentral relations according to "diencephalic" variant and to their connection with alteration of the sensitivity and bronchial responsiveness threshold in the test with metacholine.

As demonstrated by our studies, AIA patients, even before the occurrence of asthmatic syndrome, may have signs of not only disorders in the CNS and autonomic nervous system but also in the immune system, which, in particular, can manifest in the predisposition to nonmalignant tumors. Thus, 71.1% of AIA patients had polyposis rhinosinusopathy, and in 30% of them it had been diagnosed prior to the occurrence of first asthma attacks. Hysteromyoma was found in 18.8% of female patients, and 36% of female patients had indications to myomectomy and hysterectomy. The predisposition of AIA patients to multiple polyposes (of the urogenital system, intestine, large bowel, and rectum) has been noted in works by other authors [448,449]. The immune system dysfunction in AIA patients is also evidenced by our data on reduced resistance to viral infections and occurrence of chronic infectious inflammatory processes, mycetogenic, pollen, dust, food, epidermal, and drug sensibilization, as well as the onset of asthmatic syndrome after vaccination. Furthermore, we have found out that every second patient with AIA has chronic chlamydial infection caused by *Chlamydia pneumoniae* [149]. It is known that the development of virus and intracellular microorganism persistence, as well as damage of antitumor defense, is caused by immune system deficiency, including the insufficiency of T-cells and NK lymphocytes—NKs [430,634]. Some authors have pointed out to a decrease of absolute and relative numbers of T-lymphocytes, especially T-helpers, in AIA patients and to a change in their functional state, which is manifested in a disorder of T-lymphocytes response to serotonin while there is normal reaction of histamine, adrenalin, and theophylline [551]. It is believed that in AIA patients, there is insufficiency of NK cells and imbalance of cellular subpopulations sensitive to monoamines [552]. Thus, our assumptions regarding the immune system deficiency in AIA patients find support in some studies conducted by other authors.

Apart from disorders in the nervous and immune systems, we have found early manifestations of endocrine disorders. This is evidenced by the fact that 26% of female patients with AIA had menstrual cycle disorders and habitual noncarrying of pregnancy, early menopause which, in every third woman, occurred after hysterectomy. Additionally, 50% of female AIA patients of a young age have exacerbation of the asthma course in the second phase of the menstruation. The studies by Markov [336], Kagarlitskaya [250], and Suzuki et al. [553] have also demonstrated a high percentage of aspirin-sensitive asthmatics among women suffering from premenstrual bronchial asthma attacks. Besides the high-frequency reproductive function disorders, we have found that AIA patients had a higher frequency of thyroid pathology and early development of GC dependence, which points to insufficient function of the endocrine system.

The cardiovascular system of AIA patients is characterized by a neurocirculatory hypotonic distonia and early indications of the varicosity. An increased amount of platelets, disturbances in the lung microcirculation, and polyposis rhinosinusopathy of the edematous type are indicative of profound disorders in the platelet-vessel hemostasis.

The conducted mathematical analysis of indices with the use of statusmetric method allowed the discovery of, in AIA patients, certain disorders in carbohydrates metabolism due to insufficient endocrine function of the pancreas, decreased enzymatic function and increased protein-synthesizing function of the liver, damage to gastroduodenal mucous membranes, and changes in the blood-forming system with reduced proliferation of stem cells toward erythropoiesis and granulocytopoiesis with enhanced eosinophilopoiesis and megakaryocytopoiesis.

The above mentioned literature data also point to damaged AA metabolism and synthesis of eicosanoids, DNIES signaling molecules which act as local intracellular mediators.

Recent studies have demonstrated that extrapineal MT as a paracrine signaling molecule plays a key role in the regulation of AA metabolism, local coordination of cellular functions and intracellular links. At the same time, being a hormone, it takes part in integrating the activity of the endocrine and immune systems [286]. Given that "endogenic ASA" is one of MT metabolites [256], we have studied the pathogenesis of AIA from a new point, taking into account some peculiarities of MT production in this disease and its role in the homeostasis regulation.

We have found that the diurnal MT production is much lower in AIA patients compared to that in aspirin-tolerant asthmatic and healthy subjects. The change in the diurnal MT production in AIA patients is not related to the activity of NAT enzyme, which is responsible for its synthesis, but is determined by the state of DNIES cells, particularly platelets which are a main peripheral reservoir of serotonin [289,295,474]. We have shown that there is practically no MT production in platelets of AIA patients, which might be due to a damaged entry of serotonin (being a source of MT) from blood into cells. A number of researchers have pointed to decreased serotonin uptake by platelets in AIA patients [328–330].

The change in MT production in AIA patients substantially affects the platelet functional activity: in case of decreased diurnal and increased nocturnal MT production, the rate of the first phase of aggregation increases, which leads to increased ADP- and heparin-induced intensity of platelet aggregation and points to increased platelet reactivity to the inducing agent. This, in turn, may point to pathology of the membrane—receptor platelet complex in AIA patients. In this connection, we have investigated the platelet reaction to addition of exogenous MT *in vitro* in bronchial asthma patients and found a specific platelet reaction in AIA patients to *in vitro* administration of minimal doses of MT, that is, an increase of intensity and rate of the first phase of ADP-induced platelet aggregation, which is due to the opening of the same receptor-operated channels for calcium ions and/or Ca^{2+} mobilization from the intracellular stores. This mechanism, in combination with low MT content in platelets, underlies the increase of their aggregation activity. This assumption is supported by data of other researchers who have found a high calcium uptake in platelets of AIA patients even under the action of low concentration of PAF [140] and have noted a paradoxical reaction of platelets in addition to adrenalin and ADP—a significant increase of cAMP level.

It is known that the membrane–receptor complex is a universal primary regulating system which, to a large extent, depends on microviscosity of the membrane lipid layer [598]. An increased rigidity of the membrane phospholipid structure due to insufficient function of the AOS and accumulation of LPO products leads to alteration of the surface receptor exposure, affecting the platelet reactivity to the inducing agent [130,356]. Antioxidant properties of MT have been discovered recently, which has a direct effect on free radicals, acting as a "scavenger" and stimulating glutathione peroxidase [477]. An increase of this enzyme activity directly correlates with the concentration of MT in tissues [401,471]. Having the antioxidant properties, MT, on the one hand, decreases the ability of free radicals to damage the DNA structure, and on the other hand, it substantially increases fluidity and decreases microviscosity of platelet membranes [114,126]. Low MT production in AIA patients may weaken the antioxidant defense, which results in damaging not only the functions of platelet receptors but also the channels of ion permeability and lipid-dependent enzymes. This is further confirmed by data on lowered glutathione peroxidase activity and increase of oxygen free radicals production in platelets of AIA patients [330,420,511,442].

Physiological concentrations of MT affect calcium metabolism in cells by blocking L-type potential-sensitive Ca^{2+} channels, changing the level of Ca^{2+}/calmodulin in the cell and calcium content in intracellular stores [474]. A lower production of MT in platelets of AIA patients, as well as in other cells of the DNIES, leads to a disorder in cellular calcium homeostasis. A consequence of this is deep alterations in platelet form and ultrastructure found in AIA patients [9].

It should be noted that the revealed peculiarities of ADP and heparin-induced platelet aggregation in AIA patients are associated not only with a low level of basic MT production but firstly with pathology of the membrane−receptor complex of platelets themselves, which is a basis for developing a specific reaction of platelets to ASA in this disease. We have demonstrated that it is only in AIA patients that the intensity of the first phase of ADP-induced platelet aggregation decreases under the action of minimal ASA concentrations. According to literature data, ASA inhibits aggregation by lowering the second-phase intensity due to suppression of TxA_2 synthesis [189]. At the same time, ASA acetylates serine at the active center of COX [257]. It has also been proved that ASA is a strong acetylating agent which, when interacting with any proteins, acetylates the end amino groups of lysine. What is the basis for the altered reaction of platelets to ASA in AIA patients? We may assume that the acetylation of proteins of the initially damaged platelet membrane−receptor complex contributes to the suppression of Ca^{2+} release from intracellular stores and thereby to lowering the intensity of platelet aggregation and its primary wave.

It should be emphasized that it is only in AIA patients that we have found a high correlation between the changes in parameters of ADP-induced platelet aggregation after addition of ASA and after addition of MT, and the alteration of the first phase of aggregation *in vitro* after preincubation with ASA or MT did not correlate with the level of endogenous MT production. The obtained data demonstrate that at low MT production in AIA patients, it has no controlling action on the functional activity of platelets which is characterized by a high tension and low adaptive capabilities. It is known that functional activity of platelets to a certain extent depends on maturity of megakaryocytes in bone marrow [538], and it is MT that, by penetrating through the megakaryocyte membrane and accumulating in the nucleus, affects some phases of translation and transduction in these cells [126]. Disruption of MT production in bone marrow of AIA patients may be a reason of platelet immaturity and alteration of their functional activity.

One of MT metabolites is N^1-acetyl-5-methoxykynurenamine (NAMK) which is chemically similar to, and has the same properties as, ASA *in vitro* [256]. Given that in our *in vitro* experiments MT and ASA generated similar changes in ADP-induced platelet aggregation in AIA patients, we might suppose that *in vivo* NAMK ("endogenous aspirin") has the same activating effect on platelets as *in vitro* MT at a dose which is 100 times smaller than its physiological concentration, that is, platelets in AIA patients remain permanently activated. This assumption can be supported by available data on a substantial increase (by 6.5 times) in the number of spherical platelets [9], and by experimental studies having demonstrated that MT is bound with the cytoplasmic "extension" on megakaryocytes and with the platelet microtubule membrane, inducing the change of platelets from disks to spheres [320]. Platelet activation may be followed by reentry of Ca^{2+} into cells, the result of which is their quick exhaustion and death. Our data on the increased number of platelets in AIA patients may reflect a compensatory reaction in case of enhanced destruction of platelets [164]. During platelet activation, calcium concentration in cytoplasm may increase and metabolism of membrane phospholipids may be enhanced (Figure 4.1). The latter two processes are closely linked and lead to blood platelets aggregation, followed by a release reaction and formation of a wide range of biologically active substances: serotonin, histamine, histamine-releasing factor, various cationic proteins, β thromboglobulin, platelet factor 4, growth factor, regulated on activation, normal T-cell expressed and secreted (RANTES), H_2O_2, and oxygen radicals, as well as PAF and AA metabolites [115,169,227,290,300]. Most of these biologically active substances have a synergic effect on platelet activity in the human body, which leads to even higher potentiation of platelet activation [515].

The effect of biologically-active substances generated by platelets is very varied. Most mediators (histamine-releasing factor, platelet factor 4, cytokine RANTES, 12-HETE, PAF) are powerful activators and chemoattractants of eosinophils. This is confirmed by literature data on high correlations between increased contents of activated platelets and eosinophils in blood of patients with AIA [9,290]. A large quantity of activated eosinophils and mast cells were found in biopsy specimens from the bronchial tree and nasal cavity in AIA patients [532,638]. It is possible that in case of low MT formation in platelets, mast cells and eosinophils, which are sources of extrapineal MT [405,406], perform the function of delivering it to tissues.

We too have found an increased content of eosinophils in the blood of AIA patients. Activation of mast cells and eosinophils, in its turn, is followed by formation of sulfidopeptide LTs which have an expressed

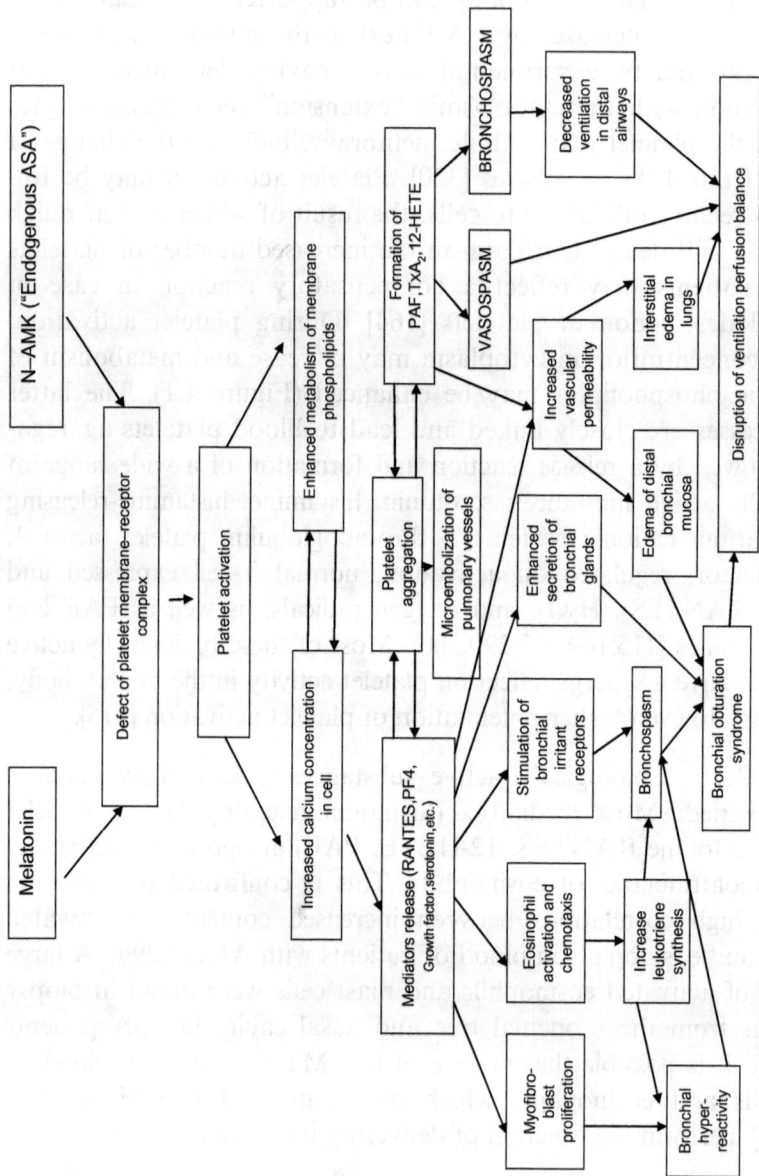

Figure 4.1 Role of platelets in pathogenesis of asthmatic syndrome in aspirin-induced asthmatics.

spasmogenic action on smooth muscles of the bronchial tree, stimulate the mucus production and vascular permeability with plasma exudation into the bronchial wall, which contributes to the development of bronchial obturation syndrome in AIA patients [91,497,581,619]. Most biologically-active substances released by platelets in AIA patients have vasoconstrictor effects, and the action of Tx *in vivo* is enhanced in the presence of other spasmogenic substances [350]. TxA_2 and PAF enhance the platelet aggregation, and the latter can also increase the vascular wall permeability for plasma components and enhance the secretion of bronchial glands [128,129]. Literature data confirm that AIA patients have especially high sensitivity of platelets to PAF [140,579], as compared to other cells, basophils in particular. The enhancement of vascular permeability is supported by excretion of cationic proteins, especially serotonin, which not only affects the lung vascular system [133,555] but also is able to stimulate irritant receptors in the bronchial epithelium and to enhance efferent parasympathetic effects. Potentiation of myofibroblast proliferation under the action of platelet growth factor [227] and serotonin, as well as the enhancement of parasympathetic tonus under the effect of the latter, may contribute to development of the bronchial tree hyperresponsiveness, which occurs in AIA patients at early stages of asthmatic syndrome formation. Furthermore, the increase of intravascular platelet aggregation results in microembolization of pulmonary vessels and, together with the effect of platelet-derived inflammatory factors, causes pulmonary microcirculation disorder, development of interstitial edema, bronchial obturation syndrome which is manifest in chronic obstructive bronchitis with expressed respiratory function disorders (Figure 4.2). This is confirmed by our data on significant airway obstruction which is mostly manifested in distal parts of the bronchial tree, and incomplete reversibility of obstruction after inhalation of β-adrenomimetic in AIA patients. The pulmonary perfusion scintigraphy has shown disturbances in the ventilation–perfusion balance in AIA patients.

As we know, thromboresistance of vessels is largely determined by biosynthesis of prostacyclin which is formed in the endothelium in response to such factors as bradykinin, angiotensin II, muscarine receptor agonists, ADP, endotoxins, chemoreceptor activators, and activated platelets [428,207]. As a vasodilatator, prostacyclin is also an antiaggregant protecting the vascular wall from platelet deposition [189,365]. Our investigations have demonstrated that prostacyclin level in blood is

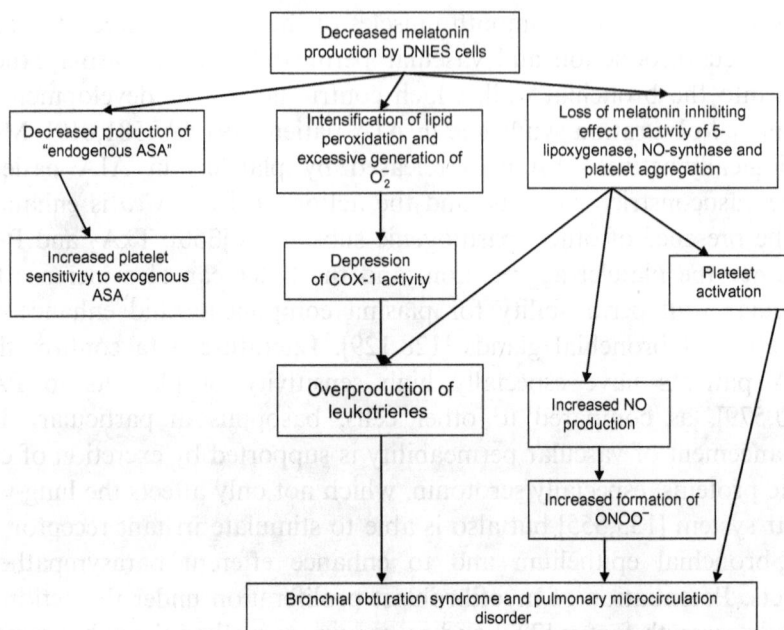

Figure 4.2 Role of MT in pathogenesis of asthmatic syndrome in patients with AIA.

lower in AIA patients than in ATA patients and healthy subjects. We believe that low MT production, not only in platelets but also in vascular endothelium, plays an important role in the genesis of lowering the prostacyclin content in plasma of AIA patients. Endothelial cells form a part of the DNIES and are a source of MT in the human body [287]. Lower MT production leads to enhancement of LPO products which are known to inhibit prostacyclin synthase, a key enzyme of prostacyclin synthesis [426,427]. Additionally, various endothelin isopeptides are formed in the endothelium, of which endothelin-1 has a strong and long vasoconstrictor action on arteries and veins [628], whereas ASA potentiates vasoconstriction caused by endothelin-1 [620]. It is possible that, due to lower prostacyclin synthesis in AIA patients, the vasoconstrictor effect of endothelin-1 is enhanced and thromboresistance of vessels is damaged, which results in expressed circulatory embarrassment and vascular tonus disorder. This is confirmed not only by pulmonary microcirculation disturbance but also by edematous polyposis rhinosinusopathy as well as by varicosity, which, according to our data, develop in AIA patients prior to the occurrence of asthmatic syndrome. Unfortunately, the role of prostacyclin in pathogenesis of AIA remains poorly studied. Researchers' attention has been only drawn to broncho-dilator action of this eicosanoid and its inhibitory effect on LT forma-tion [37,56,449]. The attempts of using synthetic preparations of prostacyclin have not had a positive effect on pulmonary function and aspirin intolerance in AIA patients [188,223,384,385]. Based on our data, we may assume that the decrease of prostacyclin production plays the most important role in the genesis of microcirculation disorders which are specific for this group of patients.

It is known that the vascular tonus and structure is also controlled by NO [482], which, additionally, has antithrombotic action, preventing platelet aggregation and initiating their disaggregation [102,206,237,334]. Interestingly, NO and prostacyclin are generated by endothelial cells simultaneously and unidirectionally both *in vitro* and *in vivo* in case of B_2 kinin, P_{2y} purinergic, and M_2 cholinergic receptors stimulation [75,206]. As our studies have demonstrated, prostacyclin synthesis is lowered and NO production is increased in patients with AIA.

The comparison of diurnal and nocturnal total urinary excretion of nitrates and nitrites has shown that nocturnal production of NO in patients with bronchial asthma and healthy subjects is higher than diurnal ones, while nocturnal NO production is much higher in AIA patients than in aspirin-tolerant asthmatics and healthy subjects. Besides, in AIA patients there is no direct correlation between noctur-nal NO production and diurnal MT production, unlike ATA patients and healthy subjects. We have not found any literature data concern-ing the circadian rhythm of NO formation in humans and its correlation with the rhythm of MT production. There are only experi-mental data that the NO content in chicken brain tissue is lower at night than at daytime, with inverse correlation between the nocturnal levels of NO and MT [210]. The increased nocturnal production of NO, as compared to diurnal one, might be caused by activation of constitutive forms of NOSs (nNOS and eNOS) in cells of the DNIES, particularly in platelets [163,148,166]. Platelets, as we know, have all three forms of NOS (neuronal, endothelial, and inducible) and gener-ate large quantities of NO—20 nmol L^{-1} of blood, with platelet con-centration equal to $1-3 \times 10^{11}$ L^{-1} [335,453,454]. Platelet pathology in AIA patients may lead not only to a decrease in MT production but also to an increase in NO generation, probably due to the loss of MT-controlling effects on the activity of cNOS and a disorder in calcium homeostasis in platelets. An increase of Ca^{2+} in cells is known to activate NOS [481].

It should be emphasized that decreased MT formation in platelets and other DNIES cells is also significant from another point of view. It is known that the enhancement of LPO in bronchial asthma leads to excessive formation of oxygen reactive radicals which suppress COX activity, which results in shunting the metabolism of AA toward increasing LT formation [184]. LTs directly activate the release of NO from polymorphonuclear granulocytes under the influence on the cellular receptor coupled to G-protein. This process is Ca^{2+} dependent [294]. On the one hand, MT is binding with the nuclear receptor RZR/RORα and suppresses the expression of 5-LO gene, a major enzyme in biosynthesis of LTs [80,582], and, on the other hand, detoxifies reactive radicals of oxygen [467,475], thereby restricting the NO formation. We believe that the decrease of MT production by DNIES cells in AIA leads to loss of its control over NO formation already at early stages of the disease. As a result, the excessive NO production in AIA patients leads to a decrease in prostacyclin synthesis in the vascular endothelium [42] and to an increased formation of peroxynitrite which is an extremely stable anion [48]. Peroxynitrite can initiate a cascade of oxidizing reaction, causing damage to proteins and lipids in cellular membranes, DNA cells [381,494,556]. As we know, peroxynitrite damages the vascular endothelium, stimulates platelet aggregation activity and levels down the inhibiting effect of NO and prostacyclin on platelet aggregation, and also reduces the vascular endothelium reaction to vasodilatator [454,536]. This results in an early disorder in peripheral circulation, reducing the vascular resistance, increasing vessel permeability and, consequently, in expressed disorders in microcirculation not only in the lung but also in other organs and tissues in AIA patients.

In contrast to AIA patients, we have found increased diurnal production of NO in aspirin-tolerant asthmatics, which conforms to the data in literature concerning high levels of exhaled NO in patient with bronchial asthma [6,11,259,505,651]. We believe that the genesis of increased NO production is based on the enhanced expression of iNOS in the epithelium of upper and lower airways under the influence of proinflammatory cytokines (TNFα, IL-1β, and IFN-γ) [235,238,379,461,479]. Given the fact that NO is a neurotransmitter of the nonadrenergic, noncholinergic (NANC) inhibitory nervous system [43,292], one may assume that the enhanced diurnal synthesis of NO in ATA patients is the human body's compensatory adaptive reaction, as NO is a powerful antioxidant [607] and, when interacting with oxygen metabolites, transforms them into less active intermediate metabolites and itself forms NO_2^- and NO_3^- ions which are excreted with urine [480,481]. Furthermore, NO inhibits the cholinergic contractile response in the trachea and main bronchi [618], which is particularly important at early stage of bronchial asthma.

Thus, the increase of NO production in aspirin-tolerant asthmatics is a reaction in response to allergen and is associated with activation of iNOS under the influence of anti-inflammatory cytokines. In aspirin-sensitive asthmatics, unlike ATA patients, the genesis of increased NO production is based on activation of cNOS in platelets, neurons, endothelial cells, and other cells in the DNIES due to the lack of MT control of its activity.

By thoroughly analyzing the obtained data and available literature, we have arrived at the conclusion that patients with AIA have a disorder in the structural and functional arrangement of those elements of the DNIES that ensure the basic endogenous production of MT by MT-producing cells. Being a major extrapineal source of MT in the human body, platelets deliver it to all organs and tissues. Thanks to its lipophilicity, MT can easily cross the histohematic barrier and cellular membranes and regulate some importance functions of the human body *in situ* [124]. The insufficiency of MT production in AIA patients is manifested in a set of objective data and subjective symptoms which constitute a certain syndrome—apudopathy. This pathology is marked by platelets combining within themselves structural and functional systems which are characteristic for glandular, phagocytic, muscle, and nervous cells. The fact that MT is formed in platelets and platelets themselves have highly sensitive acceptors to MT makes them an integral part of the DNIES which has functional and, in some organs (ganglia, epiphysis, adrenal gland), structural links with the nervous and endocrine systems. It is not without reason that some researchers view platelets as a peripheral "window of pineal MT activity" and as a pharmacological model for studying reception to neurotransmitters in the CNS [79,542]. Moreover, in our view, platelets may serve as a model for studying the state of DNIES cells. From this point of view, the platelet pathology that we have found in AIA patients, which consists in alteration of membrane–receptor complex of these cells, calcium homeostasis abnormality in them, and a distorted reaction to MT and its metabolites (NAMK), reflects a pathology in the structure and function of MT-producing cells (EC-, M-cells, etc.) in the DNIES.

As we know, apudocyte dysfunction may be caused by both innate and acquired factors, such as physical and chemical infection, oncogenous effects, acute or chronic emotional stress, occupational injury. Using the statusmetric method, we have mathematically analyzed 154 clinical and laboratory indices, which helped to determine a set of clinical and laboratory data that model a "typical image" of AIA, different as it is from ATA. It should be emphasized that we have built a high accuracy model of the AIA, which ensures accurate diagnosis (up to 92%) even without using respiratory function indices [154]. This again points to great differences in the states of all other functional systems (excluding the respiratory system) in patients with AIA and those with aspirin-tolerant bronchial asthma. The detailed analysis of indices making up the functional model of AIA points to functional insufficiency of organs in which the diffuse neuroendocrine system is represented (all parts of the gastrointestinal tract, liver, adrenals, pancreas, thymus, brain, upper airways, sympathetic ganglia, vascular endothelium), and particularly its cells which synthesize MT.

It is known that MT-producing M-cells are widely present in the upper airways [289,458], and olfactory epithelium receptors contain G_s-proteins capable of activating certain channels for Ca^{2+} [202,531]. The disorder of MT reception that we have found in AIA patients probably plays an important role in the development of recurrent nasal polyposis which is characteristic for that disease [152]. A lack of MT, which is a hormone-inhibiting cytokinesis and takes part in the regulation of cells proliferation, may contribute to tumor growth and imbalance in the synchronous interdependent activity of all DNIES cells, which is necessary for ensuring proliferation in tissues of each specific organ at a strictly determined level [18,456,457]. It is notable that in AIA patients, we have observed an inclination to tumor growth (hysteromyoma, gastric, and intestinal polyps, benign kidney lesions), which has been pointed out by other authors as well [449].

We believe that characteristic CNS disorders in AIA patients such as increased anxiety, depression, reduced adaptation to psychoemotional traumas and stress situations, link of asthma attack with weather changes, with premenstrual period, and a presence of steroid dependency of the asthma course in most of these patients at the age of 16–29 years, are directly associated with a low production of MT which has antiradical, stress-limiting and anticonvulsant properties [31,310,530]. This assumption is supported by literature data on a decrease of plasmatic MT content in patients with depression [28,107,194,299,489,626] and existence of a generalized defect in the hormonal sphere with disorder of normally interfacing relations between the epiphysis and the adrenal gland in patients with melancholic depression [26,118], and by indications to the occurrence of classical migraine episodes in 23.8% of aspirin-sensitive asthmatics long before asthma [208]. The latter fact may also be associated with low MT production, as it is known to have an analgesic effect through the impact on the central opioid system [322,643].

A dysfunction of the autonomic nervous system in AIA patients might also be associated with a low production of MT by DNIES cells, as there is a direct structural link between nerve filaments and endocrine cells of the DNIES in form of synapses, and MT itself can increase the tonus of the parasympathetic nervous system and decrease the tonus of the sympathetic nervous system [456,47,383]. Experimental studies have shown that MT increases extracellular concentration of acetylcholine [411] and inhibits the central regulation of adrenomedullary secretion of catecholamines in rats [613]. Furthermore, as we know, peptidergic nerves in the DNIES form a part of the NANC nervous system, and their main function is associated with a modulating and often inhibiting action on exciting extramural impulses [456]. Functional deficiency of the NANC system in AIA patients and imbalance between exciting and inhibiting stimuli in different parts of the autonomic nervous system may be a cause of developing hypersensitivity and hyperreactivity of the bronchial tree at the earliest stages of asthmatic syndrome formation in those patients.

It is known that MT takes part in regulating various parts of the neuroendocrine system in health and disease [85,201,275,313,375,632]. Our investigations have demonstrated that AIA progresses very quickly in patients already at a young age, and the GC dependence develops after 5 years on average. Mechanisms of steroid-dependency formation in bronchial asthma have been widely discussed in literature [526,588–590,633]. However, the GC function of the adrenals in AIA patients and the role of MT in the formation of its functional deficiency remain completely unstudied. We have shown that the activity of central parts of the HPA axis is damaged in AIA patients who have low α-MT6s urinary excretion [587]. They demonstrated an abrupt decrease of cortisol content in blood (by more than 90% of the initial background value) in response to the dexamethasone test; moreover, the decrease rate does not depend on the initial cortisol level, which, as we know, is characteristic for healthy humans and aspirin-tolerant asthmatics. Such dramatic suppression of cortisol synthesis in AIA patients in response to increase of GC hormones' level in blood after dexamethasone intake points to the suppression of the corticotropin-releasing hormone secretion by the hypothalamus and of corticotropin by the adenohypophysis.

It is known that central endocrine mechanisms are controlled by the epiphysis which provides only correctional modulation of their reaction in response to stimulation [29,31,518]. According to Arushanyan et al. [25], in the case of normal balanced functioning of the HPA, there is no need for epiphyseal correction. It can arise only in rare events of sharp deviations in the state of the HPA axis, including those provoked by stress. Under these conditions, MT may directly or indirectly interfere in the function of hypothalamus and pituitary centers, and the direction of its effects is probably determined by many factors (such as the rate of hormone level increase in blood, initial functional activity of the HPA central parts, and the epiphysis itself). The fact that a high reaction degree at the dexamethasone test in AIA patients correlated with a low diurnal initial level of MT may point to a restriction of MT damping effect on the suppression by dexamethasone of the release of hypothalamus hypophyseal hormones.

The epiphysis can control the release of these hormones by means of direct activation of MT receptors in the HPA axis [26]. However, MT receptors in pinealocytes may serve for activation of some pineal functions by the circulating hormone [26]. We may assume that a disorder of the reception to MT not only in M-cells but also in apudocytes in the endocrine glands, particularly in the HPA and in pinealocytes of the epiphysis itself in AIA patients, can lead to weakening of MT-adaptive effect on the mechanisms of homeostatic regulation in the human body. The lack of correlations between the cortisol level in plasma before and after dexamethasone intake in AIA patients may point to a damage of epiphyseal control over regulation of the HPA central parts. Under chronic stress, this may contribute to the development of secondary GC adrenal insufficiency and consequently to the steroid dependence. As continuous GC therapy may reduce MT production in the pineal [53], this leads to a "vicious circle" and promotes to even more severe course(s) of AIA.

Damaged MT production in AIA patients also affects the state of reproductive organs in which M-cells are widely represented: in the endometrium, ovary, cervical canal in females and in prostate and testicles in men [457,617]. It is know that in young women, MT production decreases in the ovulatory phase of the menstrual cycle, and also goes down during menopause, which correlates to the growth of luteotropic and follicle-stimulating pituitary hormones [49,194,476]. The decrease in MT production by the cells of the DNIES in AIA patients appears to promote the development of hormonal imbalance due to the lack of MT-adaptive control over the HPA activity. This results in a damage of the reproductive function and senilism of AIA patients. This assumption is supported by the literature data on rhythm and amplitude disturbance of MT secretion in women with menstrual disorder [331,472]. Experimental studies also have shown that MT and epiphyseal peptide extract in old female rats restore regular estrous cycles and sensitivity of their hypothalamic gender centers to estrogens, a mechanism playing a leading role in the age-dependent deprivation of reproduction. When administered to animals at the age preceding the deprivation of reproduction, these preparations significantly increase their lifetime [17,18].

The damage of MT regulatory role in the endocrine glands activity bears no hormonal function of the pancreas. This is confirmed by a disorder in the carbohydrate metabolism with hyperglycemia inclination which we revealed in AIA patients by means of statusmetric analysis. The development of these disorders in patients with AIA is probably caused not only by a low MT production by M-cells of the DNIES, which are mostly located in the gastrointestinal tract [16,26,460], but also by MT reception disorder in insulin-producing β-cells in the pancreas. This assumption may be supported by the data on existence of $Mel_{1\alpha}$ receptors, which are linked with G-proteins, on β-cells within pancreatic islets [423] and on MT influence on a circadian rhythm of insulin release by these cells [424]. It is know that the insular apparatus of the pancreas and the system providing glucose homeostasis are a target for pineal hormones, including MT which ensures not only pacemaker activity of β-cells but also adjusts the regulation of their functional activity depending on the glucose level in the blood [425,528]. Moreover, MT has a protective effect on DNA of β-cells under conditions of increased NO production and IL-1β-induced damage of β-cells [15,108], which becomes especially important in AIA in view of a sharp increase of NO production in AIA patients. The latter plays a significant role not only in the development of pancreatic insufficiency but also in damaging the gastroduodenal mucosa in AIA patients. In recent years, more attention of researchers has been drawn to disorders in "enterochromaffin MT" system rather than *Helicobacter pylori* infection [2,70,271,516,253] in the development of erosive ulcerous transformation of the ventricle and the dodecadactylon mucosa membrane. The decrease in MT production by the APUD cells in AIA patients seems to enhance LPO processes in the mucosa of the gastroduodenal tract and increase the synthesis of LTs B_4 and C_4 in tissues surrounding the ulcerous area [416] due to the lack of control by MT over the activity of 5- and 12-LO and the formation of PGs playing an anti-inflammatory role in the ventricle mucosa membrane. Having an expressed chemotaxic effect, LTB_4 contributes to leukocytes accumulation in the ventricle mucosa, secretion of oxygen radicals, lactoferrin, and hydrolytic enzymes and, in synergy with sulfidopeptide LTs, increases the small vessels permeability, which leads to edema and microcirculation disorder.

Our clinical, anamnesis and laboratory data confirm that patients with AIA tend to have a mixed pathology of the gastrointestinal tract. In this case, it is not only the gastrointestinal tract that is affected but the hepatobiliary system as well: often there is constitutional hyperbilirubinemia, lowered ALAT and ASAT activity, a higher level of total blood protein at the expense of increases in α_1-globulins, chronic acalculous or calculous cholecystitis with biliary dyskinesia. Also, according to Williams et al. [629], in AIA patients there have been noted a low activity of aspirin esterase involved in the metabolism of ASA. As we know, it is the liver where MT metabolism takes place with the help of microsomal enzymes [173]. MT and serotonin-producing cells of the APUD system are widely present in the liver and gallbladder, and MT itself has a powerful effect on the contractile function of the gallbladder and bilification [457,503,645]. We believe that the decrease of MT production by hepatic apudocytes and the damaged reception of hepatocytes to MT and even to its metabolite, "endogenic ASA," cause a decrease in the enzymatic and increase in the protein-synthesizing functions of the liver, and contribute to development of hepatobiliary pathology. At that, the decrease in the liver enzymatic function may be associated with the glucuronyltransferase, a consequence of which is an increase of indirect bilirubin level in the blood serum of those patients. This assumption is supported by data on a change of MT level in blood of patients with different hepatic pathology [21], on development of hyperplasia of MT-producing DNIES cells in the antrum mucous membrane in patients with chronic calculous cholecystitis [398], and on the ability of NSAIDs to suppress the activity of hepatic transaminases [586].

The insufficiency of MT-producing DNIES cells (apudopathy) may be related to an early development of obstructive disorders of the lung ventilation in AIA patients, which we have noted above. The diffuse neuroendocrine system is widely represented in lung tissue. For the first time, bronchial and pulmonary apudocytes were described by Feyrter in 1938 and named Kulchitsky cells [111]. They release neuropeptides (substance P, vasointestinal peptide, neurokinin Y, etc.) and provide local regulation of airway ventilation, acting on smooth muscles of the bronchial wall, and regulation of the vascular tonus and mucous secretion by the bronchial epithelium [44,297,591]. Notably, in the distal parts of the bronchial tree there are apudocytes that, by their ultrastructure, are similar to intestinal enterochromaffin cells and synthesize amines (serotonin, dopamine, MT), which, in addition to regulating the air distribution function of the bronchi, also provide the local regulation of fluid and electrolyte transport in the lung tissue [298,457]. We may assume that the hypofunction of apudocytes themselves play an essential role in the genesis of early development of the bronchial obturation syndrome in the distal parts of the bronchial tree and expressed microcirculation disorder in the lungs of AIA patients, along with pathology of the platelet-vessel hemostasis. The consequence of this is the lack of local adequate regulation of water–electrolytic exchange.

The functional insufficiency of MT-producing cells in the DNIES, as well as their paradoxical reaction to MT due to membrane–receptor pathology, seems to damage the immunological reactivity of patients with AIA. They demonstrated an early tendency for benign tumor growth in these patients, atopic sensibilization and reduced resistance to viral infections, existence of chronic infectious inflammatory processes point to initial weakening of the cell-mediated immunity, which is probably due to reduced activity and/or quantity of Th_1 lymphocytes. In the literature, there are some references to a damaged response of T-lymphocytes to serotonin in AIA patients and reduction of quantity of T-helpers and NK cells in which, as we know, MT is synthesized [456,552]. Many researchers have demonstrated MT effect on cell-mediated and humoral aspects of immunity [18,319,378,444]. The expression of G-protein-coupled receptors to MT has been found on the T-helper cell membrane [332], where T-helpers (CD4+) have the highest affinity to MT. Coupled with these receptors, MT stimulates the release of IFN-γ, IL-2, and opioid cytokines [324,325,433,468]. Furthermore, MT enhances the production of IL-1, IL-6, IL-12 by monocytes [369], increases the activity of NK cells [325], and inhibits the production of TNF-α [312] and expression of transcription nuclear factor, NF-κB, which regulates the expression of cytokines [616]. As we know, the interferon's system plays a key role in antiviral defense and has antimicrobial, antiproliferative (including antitumorogenic), immunomodulating, and other effects, determining the nonspecific resistance of the human body [186]. The most powerful producers of IFN-γ are Th_1 (CD4+) [327]. IL-1 and IL-12 enhance the synthesis of IFN-γ which, in its turn, may be deposited by NK cells, thereby increasing their cytotoxic activity [231]. A lower MT production in AIA patients by DNIES cells, including lymphocytes and NK cells, can be a reason of their functional insufficiency, decreased cytokine production, and imbalance development in the immune system and interferon's system. A consequence of this is the early formation of infection-dependent variant of the disease and early chronization of infection caused by intracellular germs *C. pneumoniae*, which we have found in most AIA patients. It is known that in case of chlamydial infection, the most critical role in recovery belongs to Th_1 helpers—their activation products being IL-2, TNF-β, and IFN-γ [66,198]. High doses of the latter have been proved to completely inhibit the growth of chlamydia. The impact of IFN-γ on epithelial cells activates the synthesis of indolamin 2,3-dioxygenase— the enzyme which stimulates an oxygen—$NADPH + H^+$-dependent phenyl-kynurenine cycle of tryptophan degradation on the mitochondrial outer membrane in cytosol. It is believed that the exhaustion of intracellular tryptophan pool may cause a chlamydial stress reaction, which leads to the formation of pathological morphological forms of chlamydia, that is, to their persistence. It is against this background that there is continuous synthesis of the heat-shock protein that plays an important role in the immunopathogenesis of persistent infection and supporting a permanent inflammatory reaction [66,198]. We may assume that the decrease of MT production in AIA patients and associated changes in their immunity and interferon systems contribute to early chronization of the process, which is confirmed by a high frequency of finding IgA antibodies to *C. pneumoniae* in patients under 50 years, and their higher titer as compared to that in ATA patients. This assumption is supported by our data on occurrence of an infection-dependent variant of asthma in 86% of AIA patients already at the age of 16−29 years, whereas in the same age group of aspirin-tolerant asthmatics the percentage is only 51.4%. It is possible that *C. pneumoniae* infection plays a significant role not only in exacerbation of asthmatic syndrome but also in substantial aggravation of the already existing disorders of the platelet-vessel hemostasis in AIA patients, because under the conditions of reduced immune resistance, this germ persists not only in epithelial cells, alveolar macrophages, and fibroblasts of infected mucous membranes but also in the vascular endothelium. Our data point to a need for timely diagnostics of chlamydial infection and differentiated approach to selection of antibacterial therapy for this kind of patients in case of disease exacerbation.

Low MT production in AIA patients leads to functional insufficiency of Th_1 and consequently to the lack of inhibiting effect of IFN-γ on the formation of Th_2. Therefore, in AIA patients with decreased antiviral immunity one may observe the activation of Th_2 and production of IL-5 by them, which is confirmed by data on enhanced expression of IL-5 in bronchial tissue of these patients [346]. IL-5 promotes the proliferation and differentiation of stimulated B-lymphocytes, activates and contributes to eosinophil maturation [139,547]. High eosinophilia of blood and mucosa, bronchial and polyposis tissue is considered a marker of AIA, which has been demonstrated by our investigations and confirmed by other authors [436,546]. The activation and chemotaxis of eosinophils in inflammation are promoted by LTs and eotaxins, which are also released in large quantities in the disease [357,443], as well as chemokines of activated platelets, and the contact of the latter with T-cells enhances the platelet activation even more [252]. It should be noted that the immune system imbalance in patients with AIA might be associated with the disturbance of MT production not only in platelets but also in other MT-producing cells in the DNIES (mast cells, NK cells, eosinophils, and endothelial cells). This disturbs antiradical defense and AA metabolism, which enhances the activation of these cells [19,266,358,498,500,621].

Low MT production and associated functional insufficiency of the immune system, and particularly Th_1 immunity, in most AIA patients may be of hereditary nature. As we know, it is in patients with the "asthmatic triad" that there is maximum occurrence of antigens HLA-B35 and HLA-DQw2, etc. [178,373,412,431,432,527], which are characterized by a relatively low content of killers–suppressors [647]. Furthermore, we have found that close relatives of 18.3% of patients with AIA suffered tuberculosis of respiratory organs, in which there is a disorder of T-cell immunity expressed in reduction of the functional activity of Th_1, in unbalance of intracellular cooperation and subpopulation of immunocompetent cells [186,593,608]. We believe that the abnormal formation of the latter in patients with AIA may be a result of low MT production in bone marrow, as MT is synthesized there in large quantities [41,577]. It has been demonstrated that it accelerates maturation of stem cells, affecting the myeloid, erythrocytic and megakaryocytic hematopoietic lineages [125,182], and enhances immune capability of cells, including lymphocytes [578].

Based on the results of our investigations, we have proposed a new concept of AIA pathogenesis, which will help better understand cellular and molecular mechanisms of this disease and take new approaches to its treatment [145].

We believe that AIA should be viewed as a separate nosological form of disease which is based on pathology of MT-producing DNIES cells, namely a defect in their membrane–receptor complex, decrease of MT production and, consequently, functional insufficiency of these cells [158,170]. A consequence of this is a paradoxical reaction to MT and its metabolite—N^1-acetyl-5-metoxykynuramine, which has a similar chemical structure and the same properties as ASA. Being synthesized by M-cells of the DNIES in different organs, MT goes into the blood where it forms specific conformational links with hemoglobin

[196] and is transferred to all organs and tissues, thereby providing regulation of each organ functioning and coordination of intracellular and intercellular processes. A decrease in basic MT production in AIA patients and damaged cell reception to MT leads to the development of pathological changes at organ and system levels. At that point, impairment of all functional systems in AIA patients occurs long before the development of asthmatic syndrome, which, to a large extent, determines its severity and quick progression of the disease and formation of GCs dependence (Figure 4.2). Another consequence of low MT production in AIA patients is the enhancement of LPO and excessive formation of highly reactive O_2 radicals, the decrease of PG production, the reduced MT inhibitory effect on the activities of 5-LO, NOS, and platelet aggregation, which causes the activation of these cells, overproduction of cysteinyl-LTs and NO [168,172,596]. A consequence of these processes is pulmonary microcirculation disorder and development of bronchial obturation syndrome even in those patients who have not taken aspirin or other NSAIDs. The reduction of basic MT production leads to decreased formation of its metabolite—endogenic ASA, which, in turn, underlies the increased sensitivity of MT-producing cells, platelets in particular, to it. As a result, even minimal doses of aspirin suppress the activity of COX-1, which leads to shunting of the already disordered metabolism of AA toward a more intensive production of LTs, followed by life-threatening attacks of asthma in AIA patients [150,156−161,171].

CONCLUSION

Aspirin-induced asthma is now viewed by Russian pulmonologists as a clinical form of bronchial asthma, whereas the European Guidelines treat it as a specific case of bronchial asthma requiring additional methods of treatment and prophylaxis. A special attention of researchers and medical practitioners to this disease is explained by its higher incidence rate, severe course, and rapid development of glucocorticoid dependence and disability of patients. Nasal, bronchial, and oral provocation testing remains the standard of aspirin-induced asthma diagnosis but are rarely used because of a danger of sudden and severe anaphylaxis. As for the pathogenesis of this disease, there is a generally accepted concept according to which aspirin and other nonsteroidal anti-inflammatory drugs (NSAIDs) trigger the shunting of arachidonic acid metabolism toward increased leukotriene production. The latter causes an expressed and constant contraction of the bronchial smooth muscle, edema, and cellular infiltration of the bronchial tree mucosa, and mucus hypersecretion. Indeed, such reactions are typical for patients with aspirin-induced asthma. However, an increase of cysteinyl-leukotrienes production is generally observed in such patients irrespective of intake of analgesics or other NSAIDs. This was one of the reasons for looking into the underlying cause of arachidonic acid dysmetabolism in that group of patients.

The author's own findings based on long-term research and the literature data presented in this book allow us to take a new approach to evaluation of the clinical picture and pathogenesis of aspirin-induced asthma. Clinical implications of the disease appear to be observed long before the occurrence of the first symptoms of discomfort in the respiratory system. First of all, dysfunctions of the endocrine and immune systems are revealed. Female patients demonstrate menstrual cycle disorders, noncarrying of pregnancy, and early menopause. Different thyroid pathologies are observed in every one of six patients. Most patients show decreased resistance to viral infections, a tendency to tumor growth (nasal polyposis, fibromyoma of the uterus, gastric and intestinal polyps, benign tumors in kidneys). Furthermore, there are certain changes in the central nervous system, such as increased emotional response to everyday stresses, constant anxiety and tension, and inclination to melancholic depression. Besides, patients may have chronic gastrointestinal pathologies (chronic gastritis, peptic ulcer, Gilbert disease, biliary duct diseases, etc.), glycometabolism disorders with a tendency for hyperglycemia, and changes in the hematopoietic system (decreased lymphopoiesis, erythropoiesis, granulocytopoiesis with increased eosinophilopoiesis and megakaryocytopoiesis).

On the part of the cardiovascular system, there occurs a neurocirculatory hypotonic dystonia, and an early onset of the varicosity is often observed. With that background, most patients appear to show upper airway pathologies: vasomotor or nonatopic rhinitis and polyposis rhinosinusitis. Usually, it is a perennial rhinitis with little effect of vasoconstrictors, and nasal polyposis becomes recurrent with no effect of polypectomy. The first clinical manifestations of the disease on the part of the respiratory system are generally observed in the course of hormonal alterations of the organism. Asthma attacks appear in the thirtieth and fortieth years in females and in the fortieth and fiftieth years in males, and during the prepubertal or pubertal periods in children. Shortly before the first asthma attacks, patients have a very brief preasthmatic period where the clinical picture is dominated by manifestation of chronic asthmatic bronchitis and signs of a changed bronchial reactivity. It should be emphasized that only in 28% of patients the first asthma attack is provoked by intake of acetylsalicylic acid or other NSAIDs. Often the onset of disease occurs after vaccination, stress, at the background of menopause or after polypectomy. In such case, patients with aspirin-induced asthma immediately demonstrate obstructive ventilation disorders in the distal bronchi. There usually is early evidence of microcirculation disorders; in every second patient, the lung scintigraphy revealed extensive areas where microcirculation was completely undetectable.

Thus, the analysis of the disease clinical findings permits to distinguish aspirin-induced asthma as a separate clinical entity which is characterized by hypofunction of the organism's vital systems and imbalance of their activity. As our research suggests, this pathology is based on a disruption of the structural and functional organization in those parts of the body's diffuse neuroimmunoendocrine system (DNIES) that provides the basic endogenous production of extrapineal melatonin—melatonin-producing cells—*platelets* in particular [156]. Melatonin plays a crucial role not only in regulation of biological rhythms but also in metabolic processes, particularly the metabolism of arachidonic acid and production of nitric oxide and its derivatives. By affecting the state of the cell's membrane—receptor complex, it provides the stability of intracellular processes, cell-to-cell cooperation, and communications between different systems. Therefore, a lack of its basic production in patients with aspirin-induced asthma is expressed in a set of objective data and subjective symptoms which form a certain syndrome—apudopathy. Blood platelets can act as a marker of this pathology, as they, along with enterochromaffin cells, produce the major portion of the extrapineal melatonin and also ensure its delivery to tissues thereby regulating the body's most important functions *in situ*. Therefore, the pathology of platelets revealed in patients with aspirin-induced asthma, which implies a defect of membrane—receptor organization of those cells, disruption of melatonin production in them, calcium homeostasis, and distorted reaction to melatonin and its metabolite ("endogenous acetylsalicylic acid"), reflects a structural and functional pathology of melatonin-producing cells of the DNIES. Such decrease of melatonin production and disruption of cell receptiveness to melatonin in patients with aspirin-induced asthma underlies the pathogenesis of this disease. The decreased production of extrapineal melatonin results in enhancement of lipid peroxidation and excessive formation of highly reactive O_2 radicals, decrease of prostaglandins production, loss of melatonin-inhibiting effect on the activity of 5-lipoxygenase, NO-synthase, and platelet aggregation. This leads to constant activation of platelets with all its consequences, increases leukotriene and nitric oxide production, which results in microcirculation disorder, especially in the lungs, and in the development of bronchial obturation syndrome even in patients who do not take aspirin or other NSAIDs. The reduction of basic melatonin production also leads to insufficient formation of its metabolite—endogenous acetylsalicylic acid—which, in turn, underlies the increased sensitivity to it, of platelets in particular. As a result, minimum doses of aspirin suppress the activity of cyclooxygenase-1, which leads to the shunting of the already disrupted metabolism of arachidonic acid toward higher leukotriene production and development of severe asthmatic states in patients with aspirin-induced asthma.

The proposed new concept of aspirin-induced asthma pathogenesis suggests new methods for treatment of this disease by means of correcting the melatonin content in the patient's organism. Further investigations into the mechanisms of aspirin-induced asthma as a pathology of melatonin-producing cells of the DNIES (platelets, especially) will help determine high-risk groups, develop preventive measures, and provide adequate therapy.

REFERENCES

[1] Abe V, Itoh MT, Miyata M, Ishikawa S, Sumi Y. Detection of melatonin, its precursors and related enzyme activities in rabbit. Exp Eye Res 1999;68:255−62.

[2] Abrahamovych I, Abrahamovych OO. The enterochromaffin-serotonin system and *Helicobacter* infection in the mechanisms of the staged development of peptic ulcer. Lik Sprava 1998;8:73−9.

[3] Adamjee J, Such YJ, Park HS, Shoi JH, Penrose JF, Lam BK, et al. Expression of 5-lipoxygenase and cyclooxygenase pathway enzymes in nasal polyps of patients with aspirin-intolerant asthma. J Pathol 2006;209:392−9.

[4] Akahoshi M, Obara K, Hirota T, Matsuda A, Hasegawa K, Takahashi N, et al. Functional promoter polymorphism in the TBX21 gene associated with aspirin-induced asthma. Hum Genet 2005;117:16−26.

[5] Akerstedt T, Folkard S. Sleep/wake regulation. In: Wetterberg L, editor. Light and biological rhythms in man. Stockholm: Pergamon Press; 1993. p. 237−46.

[6] Al-Ali MK, Howarth PH. Exhaled nitric oxide levels in exacerbations of asthma, chronic obstructive pulmonary disease and pneumonia. Saudi Med J 2001;22:249−523.

[7] Albina JE. On the expression of nitric oxide synthase by human macrophages. Why no No? J Leukoc Biol 1995;58:643−9.

[8] Aleksandrov VY. Cell reactivity and proteins. Leningrad: Nauka; 1985. 318 pp [in Russian].

[9] Alieva KM. Disorders of morphofunctional state of platelets and their correction by means of plateletpheresis in patients with allergic and aspirin-induced bronchial asthma. Synopsis of MD thesis. Moscow: 1992. 25 pp [in Russian].

[10] Allegra L, Blasi F, Centanni S, Cosentini R, Denti F, Raccanelli R, et al. Acute exacerbations of asthma in adults: role of *Chlamydia pneumoniae* infection. Eur Respir J 1994;7:165−8.

[11] Alving K, Weitzberg E, Lundberg JM. Increased amount of nitric oxide in exhaled air of asthmatics. Eur Respir J 1993;6:1368−70.

[12] Ameisen JC, Capron A, Joseph M, Maclouf J, Vorng H, Pancre V, et al. Aspirin-sensitive asthma: abnormal platelet response to drugs inducing asthmatic attacks. diagnostic and physiopathological implications. Int Arch Allergy Appl Immunol 1985;78:438−48.

[13] Ameisen JC, Capron A. Aspirin-sensitive asthma. Clin Exp Allergy 1990;20:127−9.

[14] An S, Goetzl EJ. Lipid mediators of hypersensitivity and inflammation. In: Middleton E, Reed CE, Ellis EF, editors. Allergy: principles and practice. St. Louis, MO: Mosby; 1998. p. 108−82.

[15] Andersson AK, Sandler S. Melatonin protects against streptozotocin, but not interleukin-1beta-induced damage of rodent pancreatic beta-cells. J Pineal Res 2001;30:157−65.

[16] Andrew A, Kramer B, Rawdon BB. The origin of gut and pancreatic neuroendocrine (APUD) cells—the last word? J Pathol 1998;186:117−8.

[17] Anisimov VN, Havinson VH, Morozov VG. The role of epiphyseal peptides in homeostasis regulation: 20 years of research. Achiev Mod Biol 1993;113:753−62.

[18] Anisimov VN. Melatonin: its role and clinical application. St. Petersburg: Sistema Publishers; 2007. 40 pp [in Russian].

[19] Antczak A, Montuschi P, Kharitonov S, Gorski P, Barnes PJ. Increased exhaled cysteinyl-leukotrienes and 8-isoprostane in aspirin-induced asthma. Am J Respir Crit Care Med 2002;166:301−6.

[20] Arcimowicz M, Balcerzak J, Samolinski BK. Nasal polyps is not a homogenous pathology. Lekasrski 2005;19:276−9.

[21] Arendt J. Melatonin and the mammalian pineal gland. London: Chapman & Hall; 1995. 331 pp.

[22] Ariel A, Chiang N, Arita M, Petasis NA, Serhan CN. Aspirin-triggered lipoxin A_4 and B_4 analogs block extracellular signal-regulated kinase-dependent TNF-α secretion from human T cells. J Immunol 2003;170:6266−72.

[23] Arkadieva GE, Petrischev NN. Interaction of platelets with the vessel wall in health and disease. In: Arkadieva GE, Petrischev NN, editors. Mechanisms of platelet-vascular homeostasis disorder. Leningrad; 1988. p. 3−16 [in Russian].

[24] Arm JP, O'Hickey SP, Spur BW, Lee TH. Airway responsiveness to histamine and leukotriene E_4 in subjects with aspirin-induced asthma. Am Rev Respir Dis 1989;140:148−53.

[25] Arushanyan EB, Arushanyan LG, Elbekyan KS. The role of epiphyseal adrenocortical abnormalities in correctional regulation of behavior. Achiev Physiol 1993;24:12−28 [in Russian].

[26] Arushanyan EB, Arushanyan LG. Modulatory properties of epiphyseal melatonin. Probl Endocrinol 1991;37:65−8 [in Russian]

[27] Arushanyan EB, Baturin VA, Popov AV. The hypothalamic suprachiasmatic nuclei as a regulator of the mammalian circadian system. Achiev Physiol 1988;19:67−86 [in Russian].

[28] Arushanyan EB, Chudnovsky VS. Depression and disruption of circadian rhythm. J Neuropathol Psychiatry 1988;88:126−36 [in Russian].

[29] Arushanyan EB. Epiphysis and behavior organization. Achiev Physiol 1991;22:122−41 [in Russian].

[30] Arushanyan EB. Melatonin pharmacology. Exp Clin Pharmacol 1992;55:72−7 [in Russian].

[31] Arushanyan EB. Participation of the epiphysis in anti-stress defense of the brain. Achiev Physiol 1996;27:31−49 [in Russian].

[32] Astafieva NG. Changes in functional and biochemical characteristics of platelets in case of a side effect of non-steroid anti-inflammatory drugs. Ther Arch 1990;62:55−9 [in Russian].

[33] Astafieva NG. Changes in the content of cyclic nucleotides and prostaglandins during the platelets aggregation in patients with aspirin-induced asthma. Abstracts of papers for the sixth scientific conference. Kaunas: 1986. p. 78−9 [in Russian].

[34] Astafieva NG. Investigation into the role of platelets in respiratory disease pathogenesis. The first congress for respiratory diseases: abstracts of papers. Kiev, 1990; Part 8. No. 130 [in Russian].

[35] Astafieva NG. Pathophysiologic importance of change in functional activities of platelets in allergic reactions and their pharmacologic correction. Proceedings of the international society for pathophysiology I. Moscow: May 28−June 1, 1991. p. 291.

[36] Avdogan S, Yerer MB, Goktas A. Melatonin and nitric oxide. J Pineal Res 2006;29:281−7.

[37] Bach MK, Brashler JK, Smith HW, Fitzpatrick FA, Sun FF, McGuire JC. 6,9-deepoxy-6,9-(phenylimino)-prostaglandin I_2 (U-60,257) a new inhibitor of leukotriene C and D synthesis: in vitro studies. Prostaglandins 1982;23:759−71.

[38] Bachert C, Wagenmann M, Hauser U, Rudack C. IL-5 synthesis is upregulated in human nasal polyp tissue. J Allergy Clin Immunol 1997;99:837−42.

[39] Bachert C, Wagenmann M, Rudack C, Hopken K, Hillebrandt M, Wang D, et al. The role of cytokines in infections sinusitis and nasal polyposis. Allergy 1998;53:2−13.

[40] Balabolkin II, Macharadze DSh. Aspirin-induced bronchial asthma in children. Allergology 1999;(4):29−31 [in Russian].

[41] Balmasova IP, Kvetnoy IM, Smorodinov AV. The endocrine function of apudocytes of immunocompetent organs in some forms of immune response. Bull Exp Biol Med 1983;96:78−9 [in Russian].

[42] Barker JE, Bakhle YS, Anderson J, Treasure T, Piper PJ. Reciprocal inhibition of nitric oxide and prostacyclin synthesis in human saphenous vein. Br J Pharmacol 1996;118:643−8.

[43] Barnes PJ. Neural mechanisms in asthma. Br Med Bull 1992;48:149−68.

[44] Barnes PJ. Neuropeptides in the lung: localization, function and pathophysiologic implications. Allergy Clin Immunol 1986;79(2):285−95.

[45] Barnes PJ. NO or no NO in asthma? Thorax 1996;51:218−20.

[46] Bartsch H, Bartsch C, Mecke D, Lippert TH. The relationship between the pineal gland and cancer: seasonal aspects. In: Wetterberg L, editor. Light and biological rhythms in man. Stockholm: Pergamon Press; 1993. p. 337−50.

[47] Baylis BW, Tranmer BI, Ohtaki M. Central and autonomic nervous system links to the APUD system (and their APUDomas). Semin Surg Oncol 1993;9(387−93):236.

[48] Beckman JS. Biochemistry of nitric oxide and peroxynitrite. In: Kubes P, editor. Nitric oxide: a modulator of cell−cell interactions in the microcirculation. New York, NY: R.G. Landes Company; 1995. p. 1−17.

[49] Bellipanni G, Bianchi P, Pierpaoli W, Bulian D, Ilyia E. Effects of melatonin in perimenopausal and menopausal women: a randomized and placebo controlled study. Exp Gerontol 2001;36:297−310.

[50] Belushkina NN, Grigoriev NB, Severina IS. Inhibition of human platelets aggregation by a new class of soluble guanylate cyclase activators generating nitric oxide. Biochem 1994;59:1689−97 [in Russian].

[51] Belvisi MG, Ward JK, Mitchell JA, Barnes PJ. Nitric oxide as a neurotransmitter in human airways. Arch Int Pharmacodyn 1995;329(97−110):239.

[52] Benitez-King G, Huerto-Delgadillo L, Anton-Tay F. Melatonin modifies calmodulin cell levels in MDCK and N1E-115 cell lines and inhibits phosphodiesterase activity *in vitro*. Brain Res 1991;557:289−92.

[53] Benyassi A, Schwartz C, Ducouret B, Falcon J. Glucocorticoid receptors and serotonin *N*-acetyltransferase activity in the fish pineal organ. Neuroreport 2001;12:889−92.

[54] Bertuglia S, Marchiafava PL, Colantuoni A. Melatonin prevents ischemia reperfusion injury in hamster cheek pouch microcirculation. Cardiovasc Res 1996;31:947−52.

[55] Bettahi I, Pozo D, Osuna C, Reiter RJ, Acuña-Castroviejo D, Guerrero JM. Melatonin reduces nitric oxide synthase activity in rat hypothalamus. J Pineal Res 1996;20:205−10.

[56] Bianco S, Robuschi M, Ceserani R, Gandolfi C. Effects of prostacyclin on aspecifically and specifically induced bronchoconstriction in asthmatic patients. Eur J Respir Dis Suppl 1980;106:81−7.

[57] Black PN, Scicchitano R, Jenkins CR, Blasi F, Allegra L, Wiodarczyk J, et al. Serological evidence of infection with *Chlamydia pneumoniae* is related to the severity of asthma. Eur Respir J 2000;15:254−9.

[58] Blasi F. *Chlamydia pneumoniae* in respiratory infections. In: Saikki P, editor. Proceedings of the fourth meeting of the European society for Chlamydia research. Helsinki; 2000. p. 231–234.

[59] Bochenek G, Nagraba K, Nizankowska E, Szczeklik A. A controlled study of 9alpha, 1-beta-PGF2 (a prostaglandin D2 metabolite) in plasma and urine of patients with bronchial asthma and healthy controls after aspirin challenge. J Allergy Clin Immunol 2003;111:743–9.

[60] Bochenek G, Banska K, Szabo Z, Nizankowska E, Szczeklik A. Diagnosis, prevention and treatment of aspirin-induced asthma and rhinitis. Curr Drug Targets Inflamm Allergy 2002;1:1–11.

[61] Bojkowski CJ, Arendt J, Shih MC, Markey SP. Melatonin secretion in humans assessed by measuring its metabolite, 6-sulfatoxymelatonin. Clin Chem 1987;33:13343–8.

[62] Bojkowski CJ, Arendt J. Annual changes in 6-sulphatoxymelatonin excretion in man. Acta Endocrinol 1988;117:470–6.

[63] Boman J. Diagnosis of *Chlamydia pneumoniae* infections. In: Saikki P, editor. Proceedings of the fourth meeting of the European society for *Chlamydia* research. Helsinki: 2000. p. 65.

[64] Boyce JA, Lam BK, Penrose JF, Friend DS, Parsons S, Owen WF, et al. Expression of LTC_4 synthase during the development of eosinophils *in vitro* from cord blood progenitors. Blood 1996;88:4338–47.

[65] Boyce JA. Eicosanoids in asthma, allergic inflammation, and host defense. Curr Mol Med 2008;8:335–49.

[66] Bragina EE, Orlova OE, Dmitriev GA. Some peculiarities of the *Chlamydia* life cycle. Atypical forms of existence (a review). Sex Transm Dis 1998;(1):39 [in Russian].

[67] Brainard GC. Signal transduction of light for melatonin regulation in humans. In: Stevens RG, Wilson BW, Anderson LE, editors. The melatonin hypothesis. Breast cancer and use of electric power. Richland, WA: Battelle Press; 1997. p. 268–94.

[68] Brown SM, Jampol LM. New concepts of regulation of retinal vessel tone. Arch Ophthalmol 1996;114:199–204.

[69] Brydon L, Roka F, Petit L, de Coppet P, Tissot M, Barrett P, et al. Dual signaling of human melatonin receptors via G (i2), G (i3), and G (q11) proteins. Mol Endocrinol 1999;13:2025–38.

[70] Brzozowski T, Konturek PC, Konturek SJ, Brydon L, Roka F, Petit L. The role of melatonin and L-tryptophan in prevention of acute gastric lesions induced by stress, ethanol, ischemia and aspirin. J Pineal Res 1997;23:79–89.

[71] Bubenik GA, Ayles HL, Friendship RM, Brown GM, Ball RO. Relationship between melatonin levels in plasma and gastrointestinal tissues and the incidence and severity of gastric ulcers in pigs. J Pineal Res 1998;24:62–6.

[72] Bubis M, Anis Y, Zisapel N. Enhancement by melatonin of GTP exchange and ADP ribosylation reactions. Mol Cell Endocrinol 1996;123:139–48.

[73] Bucher B, Gauer F, Pevet P, Masson-Pevet M. Vasoconstrictor effects of various melatonin analogs on the rat tail artery in the presence of phenylephrine. J Cardiovasc Pharmacol 1999;33:316–22.

[74] Busse R, Fleming I, Schini VB. The role of nitric oxide in physiology and pathophysiology. Heidelberg: Springer; 1995. p. 37–50.

[75] Busse R, Hecker M, Fleming I. Control of nitric oxide and prostacyclin synthesis in endothelial cells. Arzneimittelforschung 1994;44:392–6.

[76] Cagnacci A. Melatonin in relation to physiology in adult humans. J Pineal Res 1996;21:200–13.

[77] Cagnoli CM, Ataby C, Kharfamova E, Manew H. Melatonin protfrom singlet oxygen-induced apoptosis. J Pineal Res 1995;18:222–6.

[78] Cardinali DP, Del Zar MM, Vacas MI. The effect of melatonin in human platelets. Acta Physiol Pharmacol Ther Latinoam 1993;43:1–13.

[79] Cardinali DP, Lynch HJ, Wurtman RJ. Binding of melatonin to human and rat plasma proteins. Endocrinology 1972;91:1213–5.

[80] Carlberg C, Wiesenberg I. The orphan receptor family RZR/ROR, melatonin and 5-lipoxygenase: an unexpected relationship. J Pineal Res 1995;18:171–8.

[81] Celik G, Bavbek S, Misirligil Z, Melli M. Release of cysteinyl leukotrienes with aspirin stimulation and the effect of prostaglandin E (2) on this release from peripheral blood leucocytes in aspirin-induced asthmatic patients. Clin Exp Allergy 2001;31:1615–22.

[82] Cesbron JY, Capron A, Vargaftig BB, Lagarde M, Pincemail J, Braquet P, et al. Platelets mediate the action of diethylcarbamazine on microfilariae. Nature 1987;325(533–6):801.

[83] Chavis C, Vachier I, Chanez P, Bousquet J, Godard P. 5(S), 15(S)-Dihydroxyeicosa-tetraenoic acid and lipoxin generation in human polymorphonuclear cells: dual specificity of 5-lipoxygenase towards endogenous and exogenous precursors. J Exp Med 1996;183:1633–43.

[84] Chavis C, Vachier I, Godard P, Bousquet J, Chanez P. Lipoxins and other arachidonate derived mediators in bronchial asthma. Thorax 2000;55:38–41.

[85] Chazov EI, Isanchenkov VA. Epiphysis: its position and role in the system of neuro-endocrine regulation. Moscow: Nauka; 1974. 238 pp [in Russian].

[86] Chang JE, White A, Simon RA, Stevenson DD. Aspirin-exacerbated respiratory disease: burden of disease. Allergy Asthma Proc 2012;33:117–21.

[87] Chiang N, Arita M, Serhan CN. Anti-inflammatory circuitry: lipoxin, aspirin-triggered lipoxins and their receptor ALX. Prostaglandins Leukot Essent Fatty Acids 2005;73:163–77.

[88] Chiang N, Serhan CN, Dahlen S-E, Drazen JM, Hay DWP, Rovati GE, et al. The lipoxin receptor ALX: potent ligand-specific and stereoselective actions *in vivo*. Pharmacol Rev 2006;58:463–87.

[89] Choi J-H, Kim S-H, Bae J-S, Yu H-L, Sun C-H, Nahm D-H, et al. Lack of an association between a newly identified promoter polymorphism (-1702G > A) of the leukotriene C4 synthase gene and aspirin-intolerant asthma in a Korean population. Tohoku J Exp Med 2006;208:49–56.

[90] Chotoev ZhA. Bioenergetics of the myocardium in highland conditions. Frunze: Ilim; 1985. 183 pp [in Russian].

[91] Christie PE, Schmitz M. Sulfidopeptide leukotrienes in asthma. Atemw Lungenkrkh Jahrgang 1994;20:142–7.

[92] Christie PE, Schmitz-Schumann M, Spur BW, Lee TH. Airway responsiveness to leukotriene C_4 (LTC$_4$), leukotriene E$_4$ (LTE$_4$) and histamine in aspirin-sensitive asthmatic subjects. Eur Respir J 1993;6:1468–73.

[93] Christman BW, Christman JW, Dworski R, Blair IA, Prakash C. Prostaglandin E_2 limits arachidonic acid availability and inhibits leukotriene B_4 synthesis in rat alveolar macrophages by a nonphospholipase A_2 mechanism. J Immunol 1993;151:2096–104.

[94] Chuchalin AG, Astafieva NG, Geppe NA, Didkovsky NA. Diagnostics and treatment of aspirin-induced asthma—clinical recommendations. In: Chuchalin AG, editor. Bronchial asthma in adults. Atopic dermatitis. Moscow: Atmosfera; 2002. p. 209–29 [in Russian].

[95] Chuchalin AG, Sulakvelidze IV. A problem of aspirin-induced asthma. Ther Arch 1988;50:92–7 [in Russian].

[96] Chuchalin AG. Medicamental pneumopathy. Ther Arch 1991;63:4–11 [in Russian].

[97] Cieslic K, Zhu Y, Wu KK. Salicylate suppresses macrophage nitric-oxide synthase-2 and cyclooxygenase-2 expression by inhibiting CCAAT/enhancer-binding protein-beta binding via a common signaling pathway. J Biol Chem 2002;277:49304−10.

[98] Cieslik KA, Deng WG. Essential role of C-Rel in nitric-oxide synthase-2 transcriptional activation: time-dependent control by salicylate. Mol Pharmacol 2006;70:2004−14.

[99] Claustrat B, Brain P, Garry B, Roussel B, Sassolas G. A once-repeated study of nocturnal plasma melatonin patterns and sleep recordings in six normal young men. J Pineal Res 1986;3:301.

[100] Cook PG, Honeybourne D. Clinical aspects of Chlamydia pneumoniae infection. Presse Med 1995;24:278−82.

[101] Cook PJ, Davies P, Tunnicliffe W, Ayres JG, Honeybourne D, Wise R. Chlamydia pneumoniae and asthma. Thorax 1998;53:254−9.

[102] Cooke JP, Dzau VJD. Derangements of the nitric oxide synthase pathway, L-arginine, and cardiovascular diseases. Circulation 1997;96:379−82.

[103] Coon SL, Mazuruk R, Bernard M. The human serotonin N-acetyltransferase (EC 2.3.1.87) gene (AANAT): structure, chromosomal localization, and tissue expression. Genomics 1996;34:76−84.

[104] Corrigan C, Mallett K, Ying S, Roberts D, Parikh A, Scadding G, et al. Expression of the cysteinyl leukotriene receptors cysLT(1) and cystLT(2) in aspirin-sensitive and aspirin-tolerant chronic rhinosinusitis. J Allergy Clin Immunol 2005;115:316−22.

[105] Costa EJ, Lopes RH, Lamy-Freund MT. Permeability of pure lipid bilayers to melatonin. J Pineal Res 1995;19:123−6.

[106] Cowburn AS, Sladek K, Soja J, Adamek L, Nizankowska E, Szczeklik A, et al. Overexpression of leukotriene C_4 synthase in bronchial biopsies from patients with aspirin-intolerant asthma. J Clin Invest 1998;101:834−46.

[107] Crasson M, Kjiri K, L'Hermite-Baleriaux M, Ansseau M, Legros JJ. Serum melatonin and urinary 6-sulfatoxymelatonin in major depression. Psychoneuroendocrinology 2004;29:1−12.

[108] Csaba S, Ohahima H. DNA damage induced by peroxynitrite: subsequent biological effects. Nitric Oxide 1997;1:373−85.

[109] Cui P, Yu M, Luo Z, Dai M, Han J, Xiu R, et al. Intracellular signaling pathways involved in cell growth inhibition of human umbilical vein endothelial cells by melatonin. J Pineal Res 2008;44:107−14.

[110] Cunnane SC, Horrobin DF, Manku MS, Oka M. The vascular response to zinc varies seasonally: effect of pinealectomy and melatonin. Chronobiologia 1980;7:493−503.

[111] Cutz E. Introduction to pulmonary neuroendocrine cell system. Structure-function correlation. Microsc Res Tech 1997;37:1−3.

[112] Cuzzocrea S, Zingarelli B, Gilad E, Hake P. Protective effect of melatonin in carrageenan-induced models of local inflammation: relationship to its inhibitory effect on nitric oxide production and its peroxynitrite scavenging activity. J Pineal Res 1997;23:106−16.

[113] Daffern P, Muilenburg D, Hugli TE, Stevenson DD. Association of urinary eukotriene E4 excretion during aspirin challenges with severity of respiratory responses. J Allergy Clin Immunol 1999;104:559−64.

[114] Daniels WM, van Rensburg SJ, van Zyl JM, van der Walt BJ, Taljaard JJ. Free radical scavenging effects of melatonin and serotonin: possible mechanism. Neuroreport 1996;7:1593−6.

[115] Danilyak IG, Kogan AH, Bolevich S. The contribution of platelets to the generation of active oxygen forms by blood leukocytes, lipid peroxidation and anti-peroxidation activity in bronchial asthmatics. Pulmonology 1994;(2):43–7 [in Russian].

[116] Deigner HP, Haberkorn U, Kinscherf R. Apoptosis modulators in the therapy of neurodegenerative diseases. Exp Opin Investig Drugs 2000;9:747–64.

[117] Del Zar MM, Martinuzzo M, Cardinali DP, Carreras LO, Vacas MI. Diurnal variation in melatonin effect on adenosine triphosphate and serotonin release by human platelets. Acta Endocrinol 1990;123:453–8.

[118] Demisch K, Demisch L, Nickelsen T, Rieth R. The influence of acute and subchronic administration of various antidepressants on early morning melatonin plasma levels in healthy subjects: increases following fluvoxamine. J Neural Transm 1987;68:257–70.

[119] Deng WG, Tang ST, Tseng HP, Wu KK. Melatonin suppresses macrophage cyclooxygenase-2 and inducible nitric oxide synthase expression by inhibiting p52 acetylation and binding. Blood 2006;108:518–24.

[120] Deniz E, Colakoglu N, Sari A, Sonmez MF, Tugrul I, Oktar S, et al. Melatonin attenuates renal ischemia-reperfusion injury in nitric oxide synthase inhibited rats. Acta Histochem 2006;108:303–9.

[121] Di Bella L, Bruschi C, Gualano L. Melatonin affects on megakaryocyte membrane patch-clamp outward K+ current. Med Sci Monit 2002;8:527–31.

[122] Di Bella L, Gualano L, Rossi MT, Scalera G. Effect of the simultaneous action of melatonin and ADP in megakaryocytes in vitro. Boll Soc Ital Biol Sper 1979;55:389–93.

[123] Di Bella L, Gualano L, Rossi MT, Scalera G. The action of melatonin (MLT) on platelet metabolism in vitro. Boll Soc Ital Biol Sper 1979;55:323–6.

[124] Di Bella L, Gualano L. Key aspects of melatonin physiology: thirty years of research. Neuro Endocrinol Lett 2006;27:425–32.

[125] Di Bella L, Minuscoli GC. Serotonin/melatonin biological interrelations. International symposium on pineal hormones. Bowral, Australia: 1991. p. 29.

[126] Di Bella L, Rossi MT, Scalera G. Perspectives in pineal functions. In: Kappers JA, Pevet P, editors. The pineal gland of vertebrates including MAN. Amsterdam: Elsevier North-Holland Biomedical Press; 1979. p. 475–8.

[127] Di WL, Djahanbakhch O, Kadva A, Street C, Silman R. The pineal and extrapineal origins of 5-sulphatoxy-N-acetyl-serotonin in humans. J Pineal Res 1999;26:221–6.

[128] Dillon PK, Ritter AB, Duran WN. Vasoconstrictor effects of platelet-activating factor in the hamster cheek pouch microcirculation: dose-related relations and pathways of action. Circ Res 1988;62:722–31.

[129] Dillon PK, Duran WN. Effect of platelet-activating factor on microvascular perm-selectivity: dose-response relations and pathways of action in the hamster cheek pouch microcirculation. Circ Res 1988;62:732–40.

[130] Doni MG, Vassanelli P, Avventi GL, Bonadiman L, Meduri F. Importance of the liver in the regulation of platelet activity: effect of its functional exclusion (Eck fistula). Haemostasis 1980;9:36–42.

[131] Doolen S, Krause DN, Dubocovich L, Duckles SP. Melatonin mediates two distinct responses in vascular smooth muscle. Eur J Pharmacol 1998;345:67–9.

[132] Dubocovich ML, Markowska M. Functional MT1 and MT2 melatonin receptors in mammals. Endocrine 2005;27:101–10.

[133] Dvoretsky DP, Tkachenko BI. Hemodynamics in the lungs. Moscow: Meditsina; 1987. 288 pp [in Russian].

[134] Edenius C, Kumlin MM, Bjork T, Anggard A, Lindgren JA. Lipoxin formation in human nasal polyps and bronchial tissue. FEBS Lett 1990;272:25–8.

[135] Eicosanoids, aspirin and asthma. Szczekli A, Gryglewski RG, Vane JR, editors. New York. Basel. Hong Kong: 1998. 604 pp.

[136] Eidelshtein IA. Fundamental changes in the taxonomy of *Chlamydia* and related organisms of the order *Clamydiales*. Clin Microbiol Antimicrob Ther 1999;1:5–11 [in Russian].

[137] Ekelund U, Mellander S. Role of endothelium-derived nitric oxide in the regulation of tonus in large-bore arterial resistance vessels, arterioles and veins in cat skeletal muscle. Acta Physiol Scand 1990;40:301–9.

[138] Ekmekcioglu C. Melatonin receptors in humans: biological role and clinical relevance. Biomed Pharmacother 2006;60:97–108.

[139] Ellis AK, Keith PK. Nonallergic rhinitis with eosinophilia syndrome. Curr Allergy Asthma Rep 2006;6:215–20.

[140] Emirova AS, Tatarsky AR, Chuchalin AG. A study of platelets functional state in bronchial asthmatics. Ther Arch 1990;62:100–2 [in Russian].

[141] Eriksson L, Valtonen M, Laitinen JT, Paananen M, Kaikkonen M. Diurnal rhythm of melatonin in bovine milk: pharmacokinetics of exogenous melatonin in lactating cows and goats. Acta Vet Scand 1998;36:301–10.

[142] Etingof RN. Molecular aspects of studying receptors: some new trends. Achiev Mod Biol 1991;111:384–98 [in Russian].

[143] Evans BK, Mason R, Wilson VG. Evidence for direct vasoconstrictor activity of melatonin in pressurized segments of isolated caudal artery from juvenile rats. Naunyn Schmiedebergs Arch Pharmacol 1992;346:362–5.

[144] Evans TW, Chung KF, Rogers DF, Barnes PJ. Increased vascular permeability of the airways induced by platelet-activating factor: possible mechanisms. J Appl Physiol 1987;63:479–84.

[145] Evsyukova HV. Aspirin-sensitive asthma due to diffuse neuroendocrine system pathology. Neuroendocrinol Lett 2002;23:281–5.

[146] Evsyukova HV. Production of nitric oxide and prostacyclin in acetylsalicylic acid induced asthma. Exp Clin Cardiol 1998;3:87–9.

[147] Evsyukova HV, Amosov VI. The role of platelet pathology in lung microcirculation dysfunction in bronchial asthmatics. In: Nizovtsev VP, Kosarev VV. Proceedings of the conference on problems of experimental and clinical pulmonology. Samara: 1991. p. 78 [in Russian].

[148] Evsyukova HV, Evsyukova II. Neuro-immune endocrinology of the respiratory system. In: Paltsev MA, Kvetnoy IM, editors. A guide to neuroimmune endocrinology. 2nd ed. Moscow: OAO Izdatelstvo Meditsina; 2008. p. 256–65 [in Russian].

[149] Evsyukova HV, Fedoseev GB, Savicheva AM. Chlamydial infection and aspirin-induced bronchial asthma. Pulmonology 2002;(5):64–8 [in Russian].

[150] Evsyukova HV, Fedoseev GB. The contribution of arachidonic acid metabolites to allergic reactions. Allergology 2000;(4):21–6 [in Russian].

[151] Evsyukova HV, Kvetnoy IM, Zubzhitskaya LB, Muraya EV. Melatonin production in platelets and their functional activity in patients with aspirin-induced asthma. Clin Med 2007;85:37–41.

[152] Evsyukova HV, Kvetnoy IM, Zubzhitskaya LB, Okuneva EY. Peculiarities of melatonin production in nasal polyps in patients with aspirin triad. Arch Pathol 2008;70:33–5 [in Russian].

[153] Evsyukova HV, Petrischev NN. The influence of melatonin on the platelets aggregation in healthy subjects. Hum Physiol 1998;24:122–5 [in Russian].

[154] Evsyukova HV, Poddubsky GA, Razorenov GI. Diagnostics of aspirin-induced by means of a functional model. Clin Med 1995;73:32–5 [in Russian].

[155] Evsyukova HV. Aspirin-induced asthma. In: Fedoseev Prof. GB, editor. Bronchial asthma. St. Petersburg: Medinformagentstvo; 1996. p. 179–83 [in Russian]. [Chapter 5].

[156] Evsyukova HV. Aspirin-induced asthma. St. Petersburg: Nordmedizdat; 2010, 216 pp.

[157] Evsyukova HV. Aspirin-induced bronchial asthma (pathogenesis, diagnostics, treatment). Synopsis of doctoral dissertation. St. Petersburg, 2001. 34 pp [in Russian].

[158] Evsyukova HV. Aspirin-induced bronchial asthma. In: Fedoseev Prof. GB, Trofimov VI, editors. Bronchial asthma. St. Petersburg: Nordmedizdat; 2006. p. 93–105 [in Russian].

[159] Evsyukova HV. Aspirin-induced bronchial asthma. Med Bull 2003;(4):11–7 [in Russian].

[160] Evsyukova HV. Clinical findings, pathogenesis, diagnostics and treatment of aspirin-induced asthma. In: Fedoseev Prof. GB, editor. Allergology. Specific allergology, vol. 2. St. Petersburg: Nordmedizdat; 2001. p. 286–95 [in Russian].

[161] Evsyukova HV. Contribution of arachidonic aced metabolites to the pathogenesis of inflammation of the lungs and bronchi. In: Fedoseev Prof. GB, editor. Mechanisms of inflammation of the lungs and bronchi and anti-inflammatory therapy. St. Petersburg: Nordmedizdat; 1998. p. 489–506 [in Russian]. [Chapter 11.3].

[162] Evsyukova HV. Expression of melatonin in platelets of patients with aspirin-induced asthma. Eur J Clin Invest 2011;41:781–4.

[163] Evsyukova HV. Pathophysiological aspects of aspirin-induced bronchial asthma. In: Petrischev Prof. NN, editor. The homeostasis system. St. Petersburg: SPbGMU Publishers; 2003. p. 154–74 [in Russian].

[164] Evsyukova HV. Peculiarities of platelets functional activity in patients with aspirin-induced bronchial asthma. Clin Med 1991;69:26–9 [in Russian].

[165] Evsyukova HV. The influence of melatonin on the platelet functional activity in bronchial asthmatics. Ther Arch 1999;71:35–7 [in Russian].

[166] Evsyukova HV. The neuroendocrine system of the human lungs. Hum Physiol 2006;32:121–30 [in Russian].

[167] Evsyukova HV. The role of arachidonic acid metabolites in allergic reaction mechanisms. In: Fedoseev Prof. GB, editor. Allergology. General allergology, vol. 1. St. Petersburg: Nordmedizdat; 2001. p. 531–9 [in Russian].

[168] Evsyukova HV. The role of melatonin in pathogenesis of aspirin-sensitive asthma. Eur J Clin Invest 1999;29:563–7.

[169] Evsyukova HV. The role of platelets in the pathogenesis of aspirin-induced bronchial asthma. Synopsis of MD thesis. St. Petersburg: 1991. 19 pp [in Russian].

[170] Evsyukova HV, Fedoseev GB, Petrischev NN. Melatonin and aspirin-induced bronchial asthma. Med Acad J 2005;5:3–14.

[171] Evsyukova HV, Petrischev NN. The role of arachidonic acid metabolites in the pathogenesis of inflammation of the lungs and bronchi. Clin Med Pathol Physiol 1998;(1–2):120–9.

[172] Evsyukova HV. The role of nitric oxide, prostacyclin and melatonin in aspirin-sensitive asthma. In: Dryglewski RJ, Minuz P, editors. Nitric oxide basic research and clinical applications. IOS Press; 2001. p. 211.

[173] Facciola G, Hidestrand M, von Bahr C, Tybring G. Cytochrome P450 isoforms involved in melatonin metabolism in human liver microsomes. Eur J Clin Pharmacol 2001;56:881–8.

[174] Fahrenholz JM. Natural history and clinical features of aspirin-exacerbated respiratory disease. Clin Rev Allergy Immunol 2003;24:113–24.

[175] Fedoseev GB, Petrischev NN, Evsyukova EV. Aspirin-induced asthma (clinical findings, pathogenesis, treatment). Ther Arch 1997;69(3):64–8 [in Russian].

[176] Fedoseev GB, Petrischev NN, Evsyukova EV. The effect of melatonin on the functional activity of platelets in patients with aspirin-induced asthma. Pulmonology 1998;(3):40–4 [in Russian].

[177] Fedoseev GB, Petrischev NN, Evsyukova EV. The role of platelets in the pathogenesis of aspirin-induced bronchial asthma. Pulmonology 1992;(3):23–9 [in Russian].

[178] Fedoseev GB, Petrova MA, Totolyan AA. The significance of immunogenetic peculiarities of organism in the formation and development of bronchial asthma. Ther Arch 1991;63:14–8 [in Russian].

[179] Fedoseev GB, Trofimov VI. Bronchial asthma. St. Petersburg: Nordmedizdat; 2006. 308 pp [in Russian].

[180] Fedoseev GB, Emelyanov AV. Bronchial asthma: difficult and unsolved problems. Ther Arch 1991;63:74–8 [in Russian].

[181] Feltenmark S, Gautam N, Brunnstrom S, Griffiths W, Backman L, Edenius C, et al. Eoxins are proinflammatory arachidonic acid metabolites produced via the 15-lipoxygenase-1 pathway in human eosinophils and mast cells. Proc Natl Acad Sci USA 2008;105:680–5.

[182] Filyov LV, Petrov NS, Havinson VH, Morozov VG. The role of polypeptide factors of hypothalamus and epiphysis in regulating the functional activity of stem cells, precursors of granulomonopoesis. Bull Exp Biol Med 1984;98:481–2 [in Russian].

[183] Fisherman WE, Cochen GN. Alpha-beta adrenergic imbalance in intrinsic-intolerance rhinitis or asthma. Ann Allergy 1974;33:86–101.

[184] Folkerts G, Nijkamp FP. Viral infection and asthma: the modulatory role of nitric oxide on prostaglandin H synthase. In: Szczeklik A, Gryglewski RJ, Vane JR, editors. Eicosanoids, aspirin, and asthma. New York, NY: Marcel Dekker, Inc; 1998. p. 231–45.

[185] Fraschini F, Stankov B. Distribution of the melatonin receptor in the central nervous system of vertebrates. Kinetic parameters and signal transduction pathways. In: Wetterberg L, editor. Light and biological rhythms in man. Stockholm: Pergamon Press; 1993. p. 121–31.

[186] Freidlin IS, Totolyan AA. Immunopathological mechanisms of inflammation of the bronchi and lungs. In: Fedoseev GB, editor. Mechanisms of inflammation of the bronchi and lungs and anti-inflammatory therapy. St. Petersburg: Nordmedizdat; 1998. p. 308–62 [in Russian].

[187] Fujimura M, Ozawa S, Matsuda T. Effect of oral administration of a prostacyclin analog (OP-41483) on pulmonary function and bronchial responsiveness in stable asthmatic subjects. J Asthma 1991;28:419–24.

[188] Fujimura V, Miyake Y, Uotani R, Kanamori K, Matsuda T. Secondary release of thromboxane A2 in aerosol leukotriene C4—induced bronchoconstriction in guinea pigs. Prostaglandins 1988;35:427–35.

[189] Fuster V, Chesebro JH. Series on pharmacology in practice. Antithrombotic therapy: role of platelet-inhibitor drugs. Pharmacologic effects of platelet-inhibitor drugs (second of three parts). Mayo Clin Proc 1981;56:185–95.

[190] Garnier P, Fajac I, Dessanges JF, Dall'Ava-Santucci J, Lockhart A, Dinh-Xuan AT. Exhaled nitric oxide during acute changes of airways calibre in asthma. Eur Respir J 1996;9:1134–8.

[191] Gavrish TV. Initial immunological screening in bronchial asthma and its role in the prediction and selection of treatment. Synopsis of MD thesis. Chelyabinsk: 1989. 23 pp [in Russian].

[192] Geoffriau M, Brun J, Chazol G, Claustrat B. The physiology and pharmacology of melatonin in humans. Horm Res 1998;49:136–41.

[193] Geoffriau V, Claustrat B, Veldhuis J. Estimation of frequently sampled nocturnal melatonin production in humans by deconvolution analysis: evidence for episodic or ultradian secretion. J Pineal Res 1999;27:139–44.

[194] German SV. Human melatonin. Clin Med 1993;71:22–30 [in Russian].

[195] Gilad E, Cuzzocrea S, Zingarelli B, Salzman AL, Szabo C. Melatonin is a scavenger of peroxynitrite. Life Sci 1997;60:169–74.

[196] Gilad E, Wong HR, Zingarelli B, Virág L, O'Connor M, Salzman AL, et al. Melatonin inhibits expression of the inducible isoform of nitric oxide synthase in murine macrophages: role of inhibition of NFkappaB activation. FASEB J 1998;12:685–93.

[197] Gilad E, Zisapel N. High-affinity binding of melatonin to hemoglobin. Biochem Mol Med 1995;56:115–20.

[198] Glazkova LK, Akilov OE. Practical aspects of persistent *Chlamydia* infection. Sex Transm Infect 1999;(4):29–34 [in Russian].

[199] Global strategy for asthma management and prevention (updated 2007): Global Initiative for Asthma (GINA), <http://www.ginasthma.org>; 2007.

[200] Gold ME, Buga GM, Wood KS, Burns RE, Chaudhuri G, Ignarro LJ. Antagonistic modulatory role of magnesium and calcium on release of endothelium-derived relaxing factor and smooth muscle tone. Circ Res 1990;66:357–66.

[201] Golikov PP. Receptor mechanisms of glucocorticoid effect. Moscow: Meditsina; 1988. 288 pp [in Russian].

[202] Golubev AG. More and more transmitters, transductors, transporters and receptors. Intern Med Rev 1994;2:360–5 [in Russian].

[203] Granitov VM. Chlamydiosis. Moscow: Meditsinskaya Kniga; 2000. 90 pp [in Russian].

[204] Grattan CE. Aspirin sensitivity and urticaria. Clin Exp Dermatol 2003;28:123–7.

[205] Gryglewski RJ. Prostacyclin among prostanoids. Pharmacol Rep 2008;60:3011.

[206] Gryglewski R. Interactions between prostacyclin and nitric oxide. Newsletter 1995;2.

[207] Gryglewski RJ. Metabolites of arachidonic acid in the lung. In: Schmitz-Schumann M, Menz G, Page CP, editors. PAF, platelets, and asthma. Basel, Boston: Birkhauser Verlag; 1987. p. 179–93.

[208] Grzelewska-Rzymowska I, Bogucki A, Szmidt M, et al. Migraine in aspirin-sensitive asthmatics. Allergol Immunopathol 1985;13:13–6.

[209] Guerlotte J, Greve P, Bernard M. Hydroxyindole-O-methyltransferase in the chicken retina: immunocytochemical localization and daily rhythm of mRNA. Eur J Neurosci 1996;8:710–5.

[210] Guerrero JM, Pablos MI, Ortiz GG, Agapito MT, Reiter RJ. Nocturnal decreases in nitric oxide and cyclic GMP contents in the chick brain and their prevention by light. Neurochem Int 1996;29:417–21.

[211] Guerrero JM, Reiter RJ, Ortiz GG, Pablos MI, Sewerynek E, Chuang JI. Melatonin prevents increases in neural nitric oxide and cyclic GMP production after transient brain ischemia and reperfusion in the Mongolian gerbil (*Meriones unguiculatus*). J Pineal Res 1997;23:24–31.

[212] Gumral N, Naziroglu M, Ongel K, Beydiffi ED, Ozquner F, Sutcu R, et al. Antioxidant enzymes and melatonin levels in patients with bronchial asthma and chronic obstructive pulmonary disease during stable and exacerbation periods. Cell Biochem Funct 2009;27:276–83.

[213] Guo FH, De Raeve HR, Rice TW, Stuehr DJ, Thunnissen FB, Erzurum SC. Continuous nitric oxide synthesis by inducible nitric oxide synthase in normal human airway epithelium in vivo. Proc Natl Acad Sci USA 1995;92:7809–13.

[214] Gurin AV. The functional role of nitric oxide in the central nervous system. Achiev Physiol 1997;28:53–60 [in Russian].

[215] Gutkowski P. Niektore mechanizmy nadreaktywnosci oskrzeli. Pneum Pol 1988;56:340–4.

[216] Hahn DL, Bukstein D, Luskin A, Zeitz H. Evidence for Chlamydia pneumoniae infection in steroid-dependent asthma. Ann Allergy Asthma Immunol 1998;(80):45–9.

[217] Hahn DL, Dodge RW, Golubjatnikov R. Association of Chlamydia pneumoniae (strain TWAR) infection with wheezing, asthmatic bronchitis, and adult-onset asthma. JAMA 1991;266:225–30.

[218] Hahn DL, Golubjatnikov R. Asthma and chlamydial infection. J Fam Pract 1994;38:589–95.

[219] Hamad AM, Sutcliffe AM, Knox AJ. Aspirin-induced asthma: clinical aspects, pathogenesis and management. Drugs 2004;64:2417–32.

[220] Hamilos D, Leung DYM, Wood R, Cunningham L, Bean DK, Yasruel Z. Evidence for distinct cytokine expression in allergic versus nonallergic chronic sinusitis. J Allergy Clin Immunol 1995;96:537–44.

[221] Hampl V, Cornfield DN, Cowan NJ, Archer SL. Hypoxia potentiates nitric oxide synthesis and transiently increases cytosolic calcium levels in pulmonary artery endothelial cells. Eur Respir J 1995;8:515–22.

[222] Hardeland R, Pandi-Perumal SR, Cardinali DP. Melatonin. Int J Biochem 2006;38:313–6.

[223] Hardy C, Robinson C, Lewis RA, Tattersfield AE, Holgate ST. Airway and cardiovascular responses to inhaled prostacyclin in normal and asthmatic subjects. Am Rev Respir Dis 1985;131:18–21.

[224] Hedlund L, Lischko MM, Rollag MD, Niswender GD. Melatonin: daily cycle in plasma and cerebrospinal fluid of calves. Science 1977;155:686–7.

[225] Henderson Jr WR, Tang LO, Chu SJ, Tsao SM, Chiang GK, Jones F, et al. A role for cysteinyl leukotrienes in airway remodeling in a mouse asthma model. Am J Respir Crit Care Med 2002;165:108–16.

[226] Henderson WR. Lipid-derived and other chemical mediators of inflammation in the lung. J Allergy Clin Immunol 1987;79:41–4.

[227] Herd V, Page CP. Do platelets have a role as inflammatory cells? The handbook of immunopharmacology. Series Editor Clive Page. Immunopharmacology of Platelets. Joseph M, editor. London, New York: 1995. p. 2–20.

[228] Higashi N, Taniguchi M, Mita H, Osame M, Akiyama K. A comparative study of eicosanoid concentrations in sputum and urine in patients with aspirin-intolerant asthma. Clin Exp Allergy 2002;32:1484–90.

[229] Higashi N, Taniguchi M, Mita H, Yamaguchi H, Ono E, Akiyama K. Aspirin-intolerant asthma (AIA) assessment using the urinary biomarkers, leukotriene E4 (LTE4) and prostaglandin D2 (PGD2) metabolites. Allergol Int 2012;61:393–403.

[230] Hirata H, Arima M, Fukushima Y, Honda K, Augiyama K, Tokuhisa T, et al. Overexpression of the LTC4 synthase gene in mice reproduces human aspirin-induced asthma. Clin Exp Allergy 2011;41:1133–42.

[231] Hishino T, Koren HS, Uchida A. Natural killer activity and its regulation. Amsterdam: Excepta Medica; 1984, 487 pp.

[232] Holgate ST. Role of mast cells in the pathogenesis of asthma. Bull Eur Physiopathol Respir 1985;21:449−62.

[233] Holme G, Morley J. PAF in asthma (perspectives in asthma-3). Proceedings of a symposium. Canada: June, 1986. p. 169−89.

[234] Horrobin DF. The regulation of prostaglandin biosynthesis: negative feedback mechanisms and the selective control of formation of 1 and 2 series prostaglandins: relevance to inflammation and immunity. Med Hypotheses 1980;6:687−709.

[235] Howarth PH, Redington AE, Springall DR, Martin U, Bloom SR, Polak JM, et al. Epithelially derived endothelin and nitric oxide in asthma. Int Arch Allergy Immunol 1995;107:228−30.

[236] Huerto-Delgadillo L, Anton-Tay F, Benitez-King G. Effects of melatonin on microtubule assembly depend on hormone concentration: role melatonin as a calmodulin antagonist. J Pineal Res 1994;17:55−62.

[237] Ignarro LJ. Biological actions and properties of endothelium-derived nitric oxide formed and released from artery and vein. Circ Res 1989;65:1−21.

[238] Iijima H, Duguet A, Eum SY, Hamid Q, Eidelman DH. Nitric oxide and protein nitration are eosinophil dependent in allergen-challenged mice. Am J Respir Crit Care Med 2001;163:1233−40.

[239] Ishio S, Yamada H, Craft CM, Moriyama Y. Hydroxyindole-O-methyltransferase is another target for L-glutamate-evoked inhibition of melatonin synthesis in rat pinealocytes. Brain Res 1999;850:73−8.

[240] Itoh MT, Ishizuka B, Kuribayashi Y. Melatonin, its precursors, and synthesizing enzyme activities in the human ovary. Mol Hum Reprod 1999;5:402−8.

[241] Janssens SP, Shimouchi A, Quertermous F, Bloch DB, Bloch KD. Cloning and expression of a cDNA encoding human endothelium-derived relaxing factor/nitric oxide synthase. J Biol Chem 1992;267:14519−22.

[242] Jaschonek K, Muller CP, Faul C, Renn W. Free fatty acids and platelet prostacyclin binding. Thromb Haemost 1986;56:417.

[243] Jenkins C, Costello J, Hodge L. Systematic review of prevalence of aspirin induced asthma and its implications for clinical practice. BMJ 2004;328:434.

[244] Jinnai N, Sakagami T, Seigawa T, Kakihara M, Nakjima T, Yoshia K, et al. Polymorphisms in the prostaglandin E2 receptor subtype 2 gene confer susceptibility to aspirin-intolerant asthma: a candidate gene approach. Hum Mol Genet 2004;15:3203−17.

[245] Jockers R, Maurice P, Boutin JA, Delagrange P. Melatonin receptors, heterodimerization, signal transduction and binding sites: what's new? Br J Pharmacol 2008;154:1182−95.

[246] Joseph M, Auriault C, Capron A, Vorng H, Viens PA. A new function for platelets: IgE-dependent killing of schistosomes. Nature 1983;303:810−2.

[247] Joseph M, Capron A. Role of platelets in inflammation. In: Matsson P, Ahlstedt S, Venge P, editors. Clinical impact of the monitoring of allergic inflammation. London: Academic Press; 1991. p. 155−66.

[248] Joseph M, Gounni AS, Kusnierz JP, Vorng H, Sarfati M, Kinet JP, et al. Expression and functions of the high-affinity IgE receptor on human platelets and megakaryocyte precursors. Eur J Immunol 1997;27:2212−8.

[249] Jozsef L, Zouki C, Petasis NA, Serhan CN, Filep J. Lipoxin A_4 and aspirin-triggered 15-epi-lipoxin A_4 inhibit peroxynitrite formation, NF-kB and AP-1 activation, and IL-8 gene expression in human leukocytes. PNAS 2002;99:13266−71.

[250] Kagarlitskaya VA. Peculiarities of the course and therapy of bronchial asthma in patients with ovarian hormonal dysfunctions. Synopsis of MD thesis. St. Petersburg, 1992. 15 pp [in Russian].

[251] Karasek M, Winczyk K. Melatonin in humans. J Physiol Pharmacol 2006;57:19–39.

[252] Kasperska-Zajac A, Bogala B. Platelet function in anaphylaxis. J Investig Allergol Clin Immunol 2006;16:1–4.

[253] Kato K, Asai S, Murai I, Nagata T, Takahashi Y, Komuro S, et al. Melatonin's gastroprotective and antistress roles involve both central and peripheral effects. J Gastroenterol 2001;36:91–5.

[254] Kaur C, Sivakumar V, Lu J, Ling EA. Increased vascular permeability and nitric oxide production in response to hypoxia in the pineal gland. J Pineal Res 2007;42:338–49.

[255] Kawagishi Y, Mita H, Taniguchi M, Maruyama M, Oosaki R, Higashi N, et al. Leukotriene C4 synthase promoter polymorphism in Japanese patients with aspirin induced asthma. J Allergy Clin Immunol 2002;109:936–42.

[256] Kelly RW, Amato F, Seamark RF. N-acetyl-5-methoxy-kynurenamine, a brain metabolite of melatonin, is a potent inhibitor of prostaglandin biosynthesis. Biochem Biophys Res Commun 1984;121:372–9.

[257] Kenakin T. Drugs and receptors. An overview of the current state of knowledge. Drugs 1990;40:666–87.

[258] Kennaway DJ, Hugel HM. Melatonin binding sites: are they receptors? Mol Cell Endocrinol 1992;88:1–9.

[259] Kharitonov SA, Barne PJ, Chuchalin AG. Nitric oxide (NO) in expired air: a new test in pulmonology. Pulmonology 1997;(3):8–11 [in Russian].

[260] Kharitonov SA, Wells AU, O'Connor BJ, Cole PJ, Hansell DM, Logan-Sinclair RB, et al. Elevated levels of exhaled nitric oxide in bronchoectasis. Am J Respir Crit Care Med 1995;15:1889–93.

[261] Kharitonov SA, Yates D, Barnes PJ. Increased nitric oxide in exhaled air of normal human subjects with upper respiratory tract infections. Eur Respir J 1995;8:295–7.

[262] Kilesnikova LA, Yaga K, Hattori A, Reiter R. Melatonin content in tissues of wild and domesticated foxes *Vulpes fulvus*. J Evol Biochem Physiol 1993;29:482–6 [in Russian].

[263] Kim SH, Bac JS, Holloway JW, Lee JT, Suh CH, Nahm DH, et al. A polymorphism of MS4A2 (-109T > C) encoding the beta-chain of the high-affinity immunoglobulin E receptor (FcepsilonR1beta) is associated with a susceptibility to aspirin-intolerant asthma. Clin Exp Allergy 2006;36:877–83.

[264] Kim SH, Park HS. Genetic markers for differentiating aspirin-hypersensitivity. Yonsei Med J 2006;48:15–21.

[265] Kim SH, Yang EM, Park YM, Lee HY, Park HS. Differential contribution of the CysLTR1gene in patients with aspirin hypersensitivity. J Clin Immunol 2007;27:613–9.

[266] Kim SS, Park HS, Yoon HJ, Lee YM, Lee SK, Nahm DH. Enhanced serum neutrophil chemotactic activity was noted in both early and late asthmatic responses during lysine-aspirin bronchoprovocation test in ASA-sensitive asthmatic patients. J Korean Med Sci 2003;18:42–7.

[267] Klein DC, Coon SL, Roseboom PH, Weller JL, Bernard M, Gastel JA, et al. The melatonin rhythm-generating enzyme: molecular regulation of serotonin N-acetyltransferase in the pineal gland. Recent Prog Horm Res 1997;52: 307–57.

[268] Klein T, Nusing RM, Pfeilschfter J, Ullrich V. Selective inhibition of cyclooxygenase2. Biochem Pharmacol 1994;48:1605–10.

[269] Knowles RG. Nitric oxide synthases. Biochem Soc Trans 1996;24(3):875−8.

[270] Knox AJ. Haw prevalent is aspirin induced asthma? Thorax 2002;57:565−6.

[271] Komarov FI, Rapoport SI, Malinovskaya NK. Melatonin production if patients with duodenal ulcer at different stages of the disease. Clin Med 1998;76:15−8 [in Russian].

[272] Konig W, Bremm RD, Brom HY, Köller M, Knöller J, Raulf M, et al. The role of leukotriene-inducing and metabolizing enzymes inflammation. Int Arch Allergy Appl Immunol 1987;82:526−31.

[273] Kopin IJ, Pare MB, Axelrod J, Weissbach H. The fate of melatonin in animals. J Biol Chem 1961;236:3072.

[274] Kornblightt LI, Finocchiaro L, Molinas FC. Inhibitory effect of melatonin on platelet activation induced by collagen and arachidonic acid. J Pineal Res 1993;14:184−91.

[275] Kosykh VA, Lasukova TV, Kozlova IuG. The state of adrenal glucocorticoid function in rats after pinealectomy. Probl Endocrinol 1987;33:58−60 [in Russian].

[276] Kowalski ML, Sliwinska-Kowalska M, Igarashi Y, White MV, Sliwinska-Kowalska M, Igarashi Y, et al. Nasal secretions in response to acetylsalicylic acid. J Allergy Clin Immunol 1993;91:580−98.

[277] Kowalski ML, Borowiec M, Kurowski M, Pawliczak R. Alternative splicing of cyclooxygenase-1 gene: altered expression in leucocytes from patients with bronchial asthma and associations with aspirin-induced 15-HETE release. Allergy 2007;62:628−34.

[278] Kowalski ML, Makowska JS, Blanca M, Bavbek S, Bochenek G, Bousquet J, et al. Differential effects of aspirin and misoprostol on 15-hydroxyeicosatetraenoic acid generation by leukocytes from aspirin-sensitive asthmatic patients. J Allergy Clin Immunol 2003;112:505−12.

[279] Kowalski ML. Aspirin-sensitive rhinosinusitis and asthma. Clin Allergy Immunol 2007;19:147−75.

[280] Krasusky VK. Types of the higher nervous activity. Physiology of higher nervous activity. Part II. Moscow: Nauka; 1971. p. 180−94 [in Russian].

[281] Kroll MH, Schafer A. Biochemical mechanisms of platelet activation. Blood 1989;74:1181−95.

[282] Kroncke K-D, Fehsel K, Kolb-Bachofen V. Nitric oxide: cytotoxicity versus cytoprotection—how, why, when, and where? Nitric Oxide 1997;1:107−20.

[283] Kuehl FA, Dougherty WH, Ham EA. Interactions between prostaglandins and leukotrienes. Biochem Pharmacol 1984;33:1−5.

[284] Kumlin M, Dahlen B, Bjorck T, Zetterstrom O, Granstrom E, Dahlen S-E. Urinary excretion of leukotriene E_4 and 11-dehydro-thromboxane B_2 in response to bronchial provocations with allergen, aspirin, leukotriene D_4, and histamine in asthmatics. Am Rev Respir Dis 1992;146:96−103.

[285] Kupczyk M, Antczak A, Kuprys-Lipinska A, Kuna P. Lipoxin A generation is decreased in aspirin-sensitive patients in lysine-aspirin nasal challenge in vivo model. Allergy 2009;64:1746−52.

[286] Kvetnaya TV, Knyazkin IV. Melatonin: its role and importance in age-dependent pathology. St. Petersburg: 2003. 93 pp [in Russian].

[287] Kvetnoy IM, Manokhina RP. Electronic microscopy identification of endocrine secretory granules in endothelial cells. Bull Exp Biol Med 1986;101:116−9 [in Russian].

[288] Kvetnoy IM, Raihlin NT. Functional morphology of cells carrying out neuro-immune endocrine interactions. In: Paltsev MA, Kvetnoy IM, editors. A guide to neuro-immune endocrinology, **87**. Moscow: Meditsina; 2006. p. 33−58 [in Russian].

[289] Kvetnoy IM. Extrapineal melatonin: location and role within diffuse neuroendocrine system. Histochem J 1999;31:1–12.

[290] Laidlaw TM, Kidder MS, Bhattacharyya N, Xing W, Shen S, Milne GL, et al. Cysteinyl leukotrienes overproduction in aspirin-exacerbated respiratory disease is driven by platelet-adherent leukocytes. Blood 2012;119:3790–8.

[291] Laitinen JT, Saavedra JM. Characterization of melatonin receptors in the rat suprachiasmatic nuclei: modulation of affinity with cations and guanine nucleotides. Endocrinology 1990;126:2110–5.

[292] Lammers JW, Barnes PJ, Chung KF. Nonadrenergic, noncholinergic airway inhibitory nerves. Eur Respir J 1992;5:239–46.

[293] Lands WEM. Biochemical and cellular actions of membrane lipids. Am Rev Respir Dis 1987;136:200–4.

[294] Larfars G, Lantoine F, Devynck MA, Palmblad J, Gyllenhammar H. Activation of nitric oxide release and oxidative metabolism by leukotrienes B4, C4, and D4 in human polymorphonuclear leukocytes. Blood 1999;93:1399–405.

[295] Launay JM, Lamaitre BJ, Husson HP, Dreux C, Hartmann L, Da Prada M. Melatonin synthesis by rabbit platelets. Life Sci 1982;31:1487–94.

[296] Lechin F. Asthma, asthma medication and autonomic nervous system dysfunction. Clin Physiol 2001;21:723.

[297] Leff AR. Endogenous regulation of bronchomotor tone. Am Rev Respir Dis 1988;137:198–216.

[298] Legris GJ, Will PC, Hopfer U. Inhibition of amiloride-sensitive sodium conductance by indoleamines. Proc Natl Acad Sci USA 1982;79:2046–50.

[299] Lekanina SK. The dexamethasone test and depressive conditions. J Neuropathol Psychiatry 1986;86:1417–27 [in Russian].

[300] Lellouch-Tubiana A, Lefort J, Simon MT, Pfister A, Vargaftig BB. Eosinophil recruitment into guinea pig lungs after PAF-acether and allergen administration. Modulation by prostacyclin, platelet depletion, and selective antagonists. Am Rev Respir Dis 1988;137:948–54.

[301] Leong AJS, Matthews CD. The pineal gland as an APUD-organ. Supporting evidence and implications. Med Hypotheses 1979;5:265–74.

[302] Lev NS. The pathogenic role of nitric oxide in bronchial asthma. Russ Bull Perinat Pediatr 2000;45:48–51 [in Russian].

[303] Levoye A, Jockers R, Ayoub A, Delagrange P, Savaskan E, Guillaume JL. Are G protein-coupled receptor heterodimers of physiological relevance?—Focus on melatonin receptors. Chronobiol Int 2006;23:419–26.

[304] Lewis BA, Robin J-L. Arachidonic acid derivatives as mediators of asthma. J Allergy Clin Immunol 1985;76:259–64.

[305] Lewy AJ, Wehr TA, Goodwin FK, Newsome DA, Markey SP. Light suppresses melatonin secretion in humans. Science 1980;210:1267–9.

[306] Lewy AJ. Biochemistry and regulation of mammalian melatonin production. In: Relkin R, editor. Pineal Gland. New York, NY: Elsevier; 1983. p. 77.

[307] Li B, Zhang H, Akbar M, Kim H-Y. Negative regulation of cytosolic phospholipase A2 by melatonin in the rat pineal gland. Biochem J 2000;351:709–16.

[308] Li JT. Mechanisms of asthma. Cur Opin Pulm Med 1997;3:10–6.

[309] Lin AM, Schaad NC, Schlz PE, Coon SL, Klein DC. Pineal nitric oxide synthase: characteristics, adrenergic regulation and function. Brain Res 1994;651:160–8.

[310] Lipartiti M, Franceschini D, Lanoni R, Gusella M, Giusti P, Cagnoli CM, et al. Neuroprotective effects of melatonin. Adv Exp Med Biol 1996;398:315−21.

[311] Lissoni P, Pittalis S, Ardizzoia A, Brivio F, Barni S, Tancini G, et al. Prevention of cytokine-induced hypotension in cancer patients by the pineal hormone melatonin. Support Care Cancer 1996;4:313−6.

[312] Lissoni P. The pineal gland as a central regulator of cytokine network. Neuro Endocrinol Lett 1999;20:343−9.

[313] Lopes C, Mariano M, Markus RP. Interaction between the adrenal and the pineal gland in chronic experimental inflammation induced by BCG in MICE. Inflamm Res 2001;50:6−11.

[314] Lopez-Gonzalez MA, Guerrero JM, Rojas F, Osuna C, Delgado F. Melatonin and other antioxidants prolong the postmortem activity of the outer hair cells of the organ of Corti: its relation to the type of death. J Pineal Res 1999;27:73−7.

[315] Lotufo CM, Yamashita CE, Farsky SH, Markus RP. Melatonin effect on endothelial cells reduces vascular permeability increase induced by leukotriene B4. Eur J Pharmacol 2006;534:258−63.

[316] Lovitsky SV. Diagnostics, pathogenic role and therapy of dysfunctions of the central nervous regulation in patients with bronchial asthma. Synopsis of MD thesis. 1997. 24 pp [in Russian].

[317] Luboshitzky R, Yanai D, Shen-Orr Z. Daily and seasonal variations in the concentration of melatonin in the human pineal gland. Brain Res Bull 1998;47:271−6.

[318] Lundberg JO, Farkas-Szallasi T, Weitzberg E, Rinder J, Lidholm J, Anggaard A, et al. High nitric oxide production in human paranasal sinuses. Nature Med 1995;1:370−3.

[319] Macchi MM, Bruce JN. Human pineal physiology and functional significance of melatonin. Front Neuroendocrinol 2004;25:177−95.

[320] Maclouf J, Habib A. The analysis of eicosanoids derived from platelets. The handbook of immunopharmacology. Series Editor Clive Page. Immunopharmacology of platelets. Joseph M, editor. London, New York: 1995. p. 195−208.

[321] Madueno F, Guerro MG. Use of a mutant strain of the cyanobacterim *Synechococcus* R_2 for the determination of nitrate. Anal Biochem 1991;198:200−2.

[322] Maestroni GJ, Conti A. The melatonin-immune system-opiod network. In: Reiter RJ, Lukaszyk A, editors. Advances in pineal research, 4. London: John Libbey and Company; 1990. p. 233−41.

[323] Maestroni GJ. Melatonin as a therapeutic agent in experimental endotoxic shock. J Pineal Res 1996;20:84−9.

[324] Maestroni GJ. Melatonin, stress and the immune system. Pineal Res Rev 1989;7:268.

[325] Maestroni GJ. The immunotherapeutic potential of melatonin. Expert Opin Investig Drugs 2001;10:467−76.

[326] Mahal HS, Sharma HS, Mukherjee T. Antioxidant properties of melatonin: a pulse radiolysis study. Free Radic Biol Med 1999;26:557−65.

[327] Malashenkova IK, Tazulakhova EB, Didkovsky NA. Interferons and inductors of their synthesis. Rev Ther Arch 1998;70:35−9 [in Russian].

[328] Malmgren R, Grubbstrom J, Olsson P, Theorell H, Tornling G, Unge G. Defective serotonin (5-HT) transport mechanism in platelets from patients with endogenous and allergic asthma. Allergy 1982;37:29−39.

[329] Malmgren R, Olsson P, Tornlling G, Unge G. Acetylsalicylic asthma and migraine—a defect in serotonin (5-HT) uptake in platelets. Thromb Res 1978;13:1137−9.

[330] Malmgren R, Unge G, Zetterstrom O, Theorell H, de Wahl K. Lowered glutathione-peroxidase activity in asthmatic patients with food and aspirin intolerance. Allergy 1986;41:43−5.

[331] Maltseva LI, Gafarova EA. The role of melatonin in the development of female meno-pausal syndrome and possible applications of melatonin in treatment of abnormal climac-teric symptoms. Russ Med J 2007;15:266−9 [in Russian].

[332] Malysheva OA, Shirinsky VS. Physiological properties, pathogenic importance and clinical application of epiphyseal hormone—melatonin. Int J Immunorehab 1998;(9):80−92 [in Russian].

[333] Marchenko VN, Lototsky AY, Lovitsky SV. The role of nervous system alterations in the pathogenesis of bronchial obstruction in patients with bronchial asthma. In: Fedoseev GB, editor. Bronchial asthma. St. Petersburg: Medinformagentstvo; 1996. p. 12−6 [in Russian].

[334] Marcus AJ, Safier LB, Broekman MJ, Islam N, Fliessbach JH, Hajjar KA, et al. Thrombosis and inflammation as multicellular processes: significance of cell−cell interac-tions. Thromb Haemost 1995;74:213−7.

[335] Marek WR, Salas E. Biological significance of nitric oxide in platelet function. In: Kubes P, editor. Nitric oxide: a modulator of cell−cell interactions in the microcirculation. R.G. Landes Company; 1995. p. 43−74.

[336] Markov AE, Kozachuk IA. Clinical immunological and allergic peculiarities of aspirin-induced asthma. Med Pract 1991;982:65−7 [in Russian].

[337] Markov HM. Nitric oxide and carbonic oxide—a new class of signaling molecules. Achiev Physiol 1996;27:30−43 [in Russian].

[338] Maronde E, Middendorff R, Mayer B, Olcese J. The effect of NO-donors in bovine and rat pineal cells: stimulation of cGMP and cGMP-independent inhibition of melatonin synthe-sis. J Neuroendocrinol 1995;7:207−14.

[339] Martin FJ, Atienza G, Aldegunde M, Miguez JM. Melatonin effect on serotonin uptake and release in rat platelets: diurnal variation in responsiveness. Life Sci 1993;53:1079−87.

[340] Martin TR, Merritt TL, Raughi G, Henderson WR. Leukotriene B4 is the predominant neutrophil chemotaxin produced by the human alveolar macrophage. Am Rev Respir Dis 1985;131:37.

[341] Martin U, Bryden K, Devoy M, Howarth P. Increased levels of exhaled nitric oxide during nasal and oral breathing in subjects with seasonal rhinitis. J Allergy Clin Immunol 1996;97:768−72.

[342] Martinuzzo M, Del Zar MM, Cardinali DP, Carreras LO, Vacas MI. Melatonin effect on arachidonic acid metabolism to cyclooxygenase derivatives in human platelets. J Pineal Res 1991;11:11−5.

[343] Mascia K, Haselkorn T, Deniz YM, Miller DP, Bleecker ER, Borish L. Aspirin sensitivity and severity of asthma: evidence for irreversible airway obstruction in patients with severe or difficult-to-treat asthma. J Allergy Clin Immunol 2005;116:970−5.

[344] Massaro AF, Gaston B, Kita D, Fanta C, Stamler JS, Drazen JM. Expired nitric oxide levels during treatment of acute asthma. Am J Respir Crit Care Med 1995;1562:800−3.

[345] Mastalerz L, Sanak M, Gawlewicz-Mroczka A, Gielicz A, Cmiel A, Szczeklik A. Prostaglandin E2 systemic production in patients with asthma with and without aspirin hypersensitivity. Thorax 2008;63:27−34.

[346] Mastalerz L, Sanak M, Szczeklik A. Serum interleukin-5 in aspirin-induced asthma. Clin Exp Allergy 2001;31:1036−40.

[347] Mayansky DN. On pathogenesis of chronic inflammation. Ther Arch 1992;64:3−7 [in Russian].

[348] Mayo JC, Sainz RM, Tan DX, Hardeland R, Leon J, Rodriguez C, et al. Anti-inflammatory actions of melatonin and its metabolites, N^1-acetyl-N^2-formyl-5-methoxyky-nuramine (AFMK) and N^1-acetyl-5-methoxykuramine (AMK), in macrophages. J Neuroimmunol 2005;165:139−49.

[349] Meerson FZ, Pshennikova MG. Adaptation to stressor situation and physical loads. Moscow: Meditsina; 1988. 251 pp [in Russian].

[350] Mehta JL, Nichols WW. The potential role of thromboxane inhibitors in preventing myo-cardial ischaemic injury. Drugs 1990;40:657−65.

[351] Komarov FI, Rapoport SI, Malinovskaya NK, Anisimova VN, editors. Melatonin in health and disease. Moscow: ID Medpraktika; 2004. p. 308 [in Russian].

[352] Michel T, Feron O. Nitric oxide synthases: which, where, how, and why? J Clin Invest 1997;110:2146−52.

[353] Micheletto C, Visconti M, Tognella S, Facchini FM, Dal Negro RW. Aspirin induced asthma (AIA) with nasal polyps has the highest basal LTE4 excretion: a study vs AIA without polyps, mild topic asthma, and normal controls. Eur Ann Allergy Clin Immunol 2006;38:20−3.

[354] Micheletto C, Visconti M, Tognella S, Trevisan F, Dal Negro RW. Urinary LTE4 is higher after nasal provocation test with L-ASA in bronchial than in only nasal responders. Eur Ann Allergy Clin Immunol 2007;39:162−6.

[355] Mikhailidis DP, Barradas MA, Dandona P. Heparin-induced platelet aggregation is inhib-ited by antagonists of the thromboxane pathway. Thromb Res 1986;42:719−20.

[356] Mikhailidis DP, Jeremy JY. Editorial: platelet function—the role of essential fatty acids and eicosanoids. Prostagland Leukot Essent Fatty Acids 1989;35:187−8.

[357] Min JW, Jang AS, Park SM, Lee SH, Park SW, Park CS. Comparison of plasma eotaxin family level in aspirin-induced and aspirin-tolerant asthma. Chest 2005;128:3127−32.

[358] Mita H, Endoh S, Kudon M, Kawagishi Y, Kobayashi M, Taniguchi M, et al. Possible involvement of mast-cell activation in aspirin provocation of aspirin-induced asthma. Allergy 2001;56:1061−7.

[359] Mita H, Higashi N, Taniguchi M, Higashi A, Kawagishi Y, Akivama K. Urinary 3-bromotyrosine and 3-chlorotyrosine concentrations in asthmatic patients: lack of increase in 3-bromotyrosine concentration in urine and plasma proteins in aspirin-induced asthma after intravenous aspirin challenge. Clin Exp Allergy 2004;34:931−8.

[360] Mitchell JA, De Nucci G, Warner TD, Vane JR. Different patterns of release of endothelium-derived relaxing factor and prostacyclin. Br J Pharmacol 1992;105:485−9.

[361] Modai I, Malmgren R, Asberg M, Beving H. Circadian rhythm of serotonin transport in human platelets. Psychopharmacol 1986;88:493−5.

[362] Modai I, Malmgren R, Wettenberg L, Eneroth P, Valevski A, Asberg M. Blood levels of melatonin, serotonin, cortisol, and prolactin in relation to the circadian rhythm of platelet serotonin uptake. Psychiatry Res 1992;43:161−6.

[363] Moncada S, Higgs A. The L-arginine-nitric oxide pathway. N Engl J Med 1993;329:2000−12.

[364] Moncada S, Palmer RM, Higgs EA. Nitric oxide: physiology, pathophysiology and phar-macology. Pharmacol Rev 1991;43:109−42.

[365] Moncada S. Prostacyclin and arterial wall biology. Arteriosclerosis 1982;2:143−208.

[366] Moneret-Vautrin DA, Wayoff W, Hsich V. Le Nares, maillon evolutio'e la triade de Fernand-Widal. Ann Oto-Laringol 1989;106:25−7.

[367] Monroe KK, Watts SW. The vascular reactivity of melatonin. Gen Pharmacol 1998;30:31−5.

[368] Morera AL, Abreu P. Existence of melatonin in human platelets. J Pineal Res 2005;39:432−3.

[369] Morrey KM, McLachlan JA, Serkin CD, Bakouche O. Activation of human monocytes by the pineal hormone melatonin. J Immunol 1994;153:2671−80.

[370] Morwood K, Gillis D, Smith W, Kette F. Aspirin-sensitive asthma. Intern Med J 2005;35:240−6.

[371] Muck AO, Seeger H, Bartsch C, Lippert TH. Does melatonin affect calcium influx in human aortic smooth muscle cells and estradiol-mediated calcium antagoNISM?. J Pineal Res 1996;20:145−7.

[372] Muino JC, Garnego R, Caillet Bois R, Gregorio MJ, Ferrero M, Romero-Piffiguer M. Study of cellular inflammatory response with bronchoalveolar lavage in allergic asthma, aspirin asthma and in extrinsic infiltrating alveolitis. Rev Fac Cien Med Univ Nac Cordoba 2002;59:71−82.

[373] Mullarkey S, Thomas PS, Hansen JA, Webb DR, Nisperos B. Association of aspirin-sensitive asthma with HLA-DQw2. Am Rev Respir Dis 1986;133:261−3.

[374] Nathan C. Inducible nitric oxide synthase: what difference does it make? J Clin Invest 1997;100:2417−23.

[375] Naumenko EV, Popova NK. Serotonin and melatonin in regulation of the endocrine system. Novosibirsk: Nauka; 1975. 218 pp [in Russian].

[376] Neijens HJ. Determinants and regulating processes in bronchial hyperreactivity. Lung 1990;168:268−77.

[377] Nelson CS, Marino JL, Allen CN. Melatonin receptors activate heteromeric G-protein coupled Kir3 channels. Neuroreport 1996;7:717−20.

[378] Nelson RJ, Drazen DL. Melatonin mediates seasonal changes in immune function. Ann N Y Acad Sci 2000;917:404−15.

[379] Nevzorova VA, Eliseeva EV, Zuga MV. Nitric oxidergic mechanisms of bronchi regulation and their role in bronchial asthma pathogenesis. Ther Arch 1998;70:13−8 [in Russian].

[380] Nevzorova VA, Geltser BI. Nitric oxidergic mechanisms of bronchi regulation in bronchial asthma and chronic bronchitis. In: Chuchalin AG, editor. Current problems of pulmonology. Moscow; 2000. p. 711−21 [in Russian].

[381] Nguyen T, Brunson D, Crespi CL, Penman BW, Wishnok JS, Tannenbaum SR. DNA damage and mutation in human cells exposed to nitric oxide in vitro. Proc Natl Acad Sci USA 1992;89:3030−4.

[382] Nijkamp FP, Folkerts G. Nitric oxide and bronchial reactivity. Clin Exp Allergy 1994;24:905−14.

[383] Nishiyama K, Yasue H, Moriyama Y, Tsunoda R, Ogawa H, Yoshimura M, et al. Acute effects of melatonin administration on cardiovascular autonomic regulation in healthy men. Am Heart J 2001;141:E9.

[384] Nizankovska E, Sheridan AO, Maile MH. Pharmacological attempts to modulate leukotriene synthesis in aspirin-induced asthma. In: Schmitz-Schumann M, Menz G, Page CP, editors. PAF, platelets, and asthma, 203. Basel, Boston: Birkhauser Verlag; 1987.

[385] Nizankowska E, Czerniawska-Mysik G, Szczeklik A. Lack of effect of i.v. prostacyclin on aspirin-induced asthma. Eur J Respir Dis 1986;69:363−8.

[386] Nizankowska E, Duplaga M, Bochenek G, Szczeklik A. Clinical course of aspirin-induced asthma. Results of AIANE. In: Szczklik A, Gryglewski RJ, Vane JR, editors. Eicosanoids, aspirin, and asthma. New York, NY: Marcel Dekker, Inc.; 1998. p. 451–72.

[387] Noda Y, Mori A, Liburdy R, Packer L. Melatonin and its precursors exhibit nitric oxide scavenging activity. Pathophysiology 1998;5:85.

[388] Noda Y, Mori A, Liburdy R, Packer L. Melatonin and its precursors scavenge nitric oxide. J Pineal Res 1999;27:159–63.

[389] Nosjean O, Ferro M, Coge F, Beauverger P, Henlin JM, Lefoulon F, et al. Identification of the melatonin-binding site MT_3 as the quinone reductase 2. J Biol Chem 2000;275:31311–7.

[390] Nussler Geller DA, Lowenstein CJ, Shapiro RA, Nussler AK, Di Silvio M, Wang SC, et al. Molecular cloning and expression of inducible nitric oxide synthase from human hepatocytes. Proc Natl Acad Sci USA 1993;90:3491–5.

[391] O'Sullivan S, Dahlen B, Dahlen S-E, Kumlin M. Increased urinary excretion of the prostaglandin D_2 metabolite 9α, 11β-prostaglandin F_2 after aspirin challenge supports mast cell activation in aspirin-induced airway obstruction. J Allergy Clin Immunol 1996;98:12.

[392] O'Byrne PM. Leukotrienes in the pathogenesis of asthma. Chest 1997;11:27–34.

[393] Odyvanova LG, Sosunov AA, Gatchev Ya, Cervos-Navarro J. Nitric oxide in the nervous system. Achiev Mod Biol 1997;117:374–89 [in Russian].

[394] Okada M, Sagawa T, Tominaga A, Kodama T, Hitsumoto Y. The mechanisms for platelet-mediated killing of tumour cells: one cyclo-oxygenase dependent and the other nitric oxide dependent. Immunol 1996;89:158–64.

[395] Okatani Y, Wakatsuki A, Morioka N, Watanable K. Melatonin inhibits the vasorelaxant action of peroxynitrite in human umbilical artery. J Pineal Res 1999;27:111–5.

[396] Okatani Y, Watanabe K, Hayashi K, Wakatsuki A, Sagara Y. Melatonin inhibits vasospastic action of hydrogen peroxide in human. J Pineal Res 1997;22:163–8.

[397] Oosaki R, Mizushima Y, Kawasaki A, Kobayashi M, Mita H, Maeda Y, et al. Fundamental studies on the measurement of urinary leukotriene E_4. Arerugi 1994;43:127–33.

[398] Osadchuk MA, Geraskina TB. Chronic cholecystitis: some aspects of lithogenesis. Ther Arch 1997;69:27–30 [in Russian].

[399] Ouyang H, Vogel HJ. Melatonin and serotonin interactions with calmodulin: NMR, spectroscopic and biochemical studies. Biochim Biophys Acta 1998;1383:37–47.

[400] Ozaki Y, Lyncy HJ. Presence of melatonin in plasma and urine or pinealectomized rats. Endocrinology 1976;99:641–4.

[401] Pablos MI, Chuang J, Reiter RJ, Ortiz GG, Daniels WM, Sewerynek E, et al. Time course of melatonin-induced increase in glutathione peroxidase activity in chick tissues. Biol Signals 1995;4:325–30.

[402] Palikhe NS, Kim S-H, Park HS. What do we know about the genetics of aspirin intolerance? J Clin Pharm Ther 2008;33:465–72.

[403] Palma-Carlos ML, Santos MC, Palma-Carlos AC. Agregation plaquettaire dans l'asthme. Allergie et Immunologie 1989;21:177–82.

[404] Palmer RM, Ashton DS, Moncada S. Vascular endothelial cells synthesise nitric oxide from L-arginine. Nature 1988;328:664–6.

[405] Paltsev MA, Kvetnoy IM. A guide to neuro-immune endocrinology. 2nd ed. Moscow: OAO Izdatelstvo Meditsina; 2008. 512 pp [in Russian].

[406] Paltsev MA, Kvetnoy IM. A guide to neuro-immune endocrinology. Moscow: OAO Izdatelstvo Meditsina; 2006. 384 pp [in Russian].

[407] Pandi-Perumal SR, Srinivasan V, Maestroni GJM, Cardinali DP, Poeggeler B, Hardeland R. Melatonin. Nature's most versatile biological signal? FEBS J 2006;273:2813–38.

[408] Pandi-Perumal SR, Trakht I, Srinivasan V, Spence DW, Maestroni GJ, Zisapel N, et al. Physiological effects of melatonin receptors and signal transduction pathways. Prog Neurobiol 2008;85:335–53.

[409] Pang SF, Dubocovich ML, Brown GM. Melatonin receptors in peripheral tissues: a new area of melatonin research. Biol Signals 1993;2:177–80.

[410] Pappolla MA, Chyan Y-J, Bozner P. Dual anti-amyloidogenic and antioxidant properties of melatonin: a new therapy for alzheimer's disease?. In: Iqbal K, Swaab DF, Winblad B, Wisniewski HM, editors. Alzheimer's disease and related disorders. John Wiley & Sons; 1999. p. 661–9.

[411] Paredes D, Rada P, Bonilla E, Gonzalez LE, Parada M, Hernandez L. Melatonin acts on the nucleus accumbens to increase acetylcholine release and modify the motor activity pattern of rats. Brain Res 1999;85:14–20.

[412] Park BL, Kim TH, Kim JH, Bae JS, Pasaje CF, Cheong HS, et al. Genome-wide association study of aspirin-exacerbated respiratory disease in a Korean population. Hum Genet 2013;132:313–21.

[413] Park HT, Baek SY, Kim BS, Kim JB, Kim JJ. Developmental expression of RZR beta, a putative nuclear-melatonin receptor mRNA in the suprachiasmatic nucleus of the rat. Neurosci Lett 1996;217:17–20.

[414] Parker ChV. Mediators: release and functions. In: Paul W, editor. Immunology. Moscow: Mir; 1989, chapter 27. p. 213–31 [in Russian].

[415] Pasaje CF, Bae JS, Park BL, Cheong HS, Kim JH, Jang AS, et al. DCBLD2 gene variations correlate with nasal polyposis in Korean asthma patients. Lung 2012;190:199–207.

[416] Pasechnikov VD. Synthesis of leukotrienes B4 and C4 in the ulcerous gastric mucosa. Ther Arch 1991;63:16–8 [in Russian].

[417] Petrischev NN, editor. Pathophysiology of microcirculation and hemostasis. St. Petersburg: CPbGMU; 1998. p. 500 [in Russian].

[418] Pavord ID, Tattersfield AE. Bronchoprotective role for endogenous PGE$_2$. Lancet 1995;345:436–8.

[419] Pearse AGE. Common cytochemical and ultrastructural characteristics of cells producing polypeptide hormones (the APUD series) and their relevance to thyroid and ultimobranchial C-cells and calcitonin. Proc Roy Soc Lond B Biol Sci 1968;170:71–80.

[420] Pearson J, Suarez-Mendez J. Abnormal platelet hydrogen peroxide metabolism in aspirin hypersensitivity. Clin Exp Allergy 1990;20:157–63.

[421] Penny R. Episodic secretion of melatonin in pre- and postpubertal girls and boys. J Clin Endocrinol Metab 1985;60:751.

[422] Perez-Novo CA, Watelet JB, Claeys C, Van Cauwenberge P, Bachert C. Prostaglandin, leukotriene and lipoxin balance in chronic rhinosinusitis with and without nasal polyps. J Allergy Clin Immunol 2005;115:1189–96.

[423] Peschke E, Fauteck J-D, Mubhoff U, Schmidt F, Beckmann A, Peschke D. Evidence for a melatonin receptor within pancreatic islets of neonate rats: functional, autoradiographic and molecular investigations. J Pineal Res 2000;28:156–64.

[424] Peschke E, Peschke D. Evidence for a circadian rhythm of insulin release from perifused rat pancreatic islets. Diabetologia 1998;41:1085–92.

[425] Peschke E, Peschke D, Hammer T, Csemus V. Influence of melatonin and serotonin on glucose-stimulated insulin release from perifused rat pancreatic islets *in vitro*. J Pineal Res 1997;23:156−63.

[426] Petrischev NN, Fedoseev GB, Evsyukova EV. Influence of platelets activity dysfunctions on microcirculation in the lungs and respiratory function in patients with aspirin-induced asthma. Ther Arch 1993;65:12−5 [in Russian].

[427] Petrischev NN. Thrombogenic properties of vessels. In: Petrischev NN, editor. Pathological physiology of hemostasis system. Leningrad: 1990. p. 3−24 [in Russian].

[428] Petrischev NN. Thromboresistance of vessels. St. Petersburg: ANT; 1994. 130 pp. [in Russian].

[429] Petrishchev NN, Vlasov TD. Mesenterial microcirculation and postischemic reperfusion of rat brain. Pathophysiology 2000;6:271−4.

[430] Petrov RV. Immunology. Moscow: Meditsina; 1982. 368 pp [in Russian].

[431] Petrova MA. Analysis of some risks of occurrence and development of bronchial asthma and clinical pathogenic variants of its formation. Synopsis of MD thesis. Leningrad: 1986. 17 pp [in Russian].

[432] Petrova MA. Clinical pathogenic "portraits" of bronchial asthmatic patients, carriers of certain HLA-antigens. Proceedings 2001;8:32−4 [in Russian].

[433] Petrovsky N, Harrison LC. Diurnal rhythmicity of human cytokine production: a dynamic disequilibrium in T helper cell type1/T helper cell type 2 balance? J Immunol 1997;158:5163−8.

[434] Pfaar O, Klimek L. Eicosanoids, aspirin-intolerance and the upper airways-current standards and resent improvements of the desensitization therapy. J Physiol Pharmacol 2006;57:5−13.

[435] Picado C, Fernandez-Morata GC, Juan M, Roca-Ferrer J, Fuentes M, Xaubet A, et al. Cyclooxygenase-2 mRNA is down-expressed in nasal polyps from aspirin-sensitive asthmatics. Am J Respir Cell Mol Biol 2000;160:291−6.

[436] Picado C. Aspirin intolerance and nasal polyposis. Curr Allergy Asthma Rep 2002;2:488−93.

[437] Pierzchalska M, Soja J, Wos M, Szabo Z, Nizankowska-Mogielnicka E, Sanak M, et al. Deficiency of cyclooxygenases transcripts in cultured primary bronchial epithelial cells of aspirin-sensitive asthmatics. J Physiol Pharmacol 2007;58:207−18.

[438] Pierzchalska M, Szabo Z, Sanak M, Soja J, Szczeklik A. Deficient prostaglandin E2 production by bronchial fibroblasts of asthmatic patients, with special reference to aspirin-induced asthma. J Allergy Clin Immunol 2003;111:1041−8.

[439] Pintor J, Martin L, Pelaez T, Hoyle CH, Peral A. Involvement of melatonin MT (3) receptors in the regulation of intraocular pressure in rabbits. Eur J Pharmacol 2001;416:251−4.

[440] Planaguma A, Kazani S, Marigowda G, Haworth O, Maiani TJ, Israel E, et al. Airway lipoxin A4 generation and lipoxin A4 receptor expression are decreased in severe asthma. Am J Respir Crit Care Med 2008;178:574−82.

[441] Planaguma A, Titos E, Lopez-Parra M, Gaya J, Pueyo G, Arroyo V, et al. Aspirin (ASA) regulates 5-lipoxygenase activity and peroxisome proliferator-activated receptor α-mediated CINC-1 release in rat liver cells: novel actions of lipoxin A4(LXA4) and ASA-triggered 15-epi- LXA4. FASEB J 2002;16:1937−9.

[442] Plaza V, Prat J, Roselló J, Ballester E, Ramis I, Mullol J, et al. *In vitro* release of arachidonic acid metabolites, glutathione peroxidase, and oxygen-free radicals from platelets of asthmatic patients with and without aspirin intolerance. Thorax 1995;50:490−6.

[443] Pods R, Ross D, van Hulst S, Rudack C, Maune S. RANTES, eotaxin and eotaxin-2 expression and production in patients with aspirin triad. Allergy 2003;58:1165–70.

[444] Poon AM. Evidence for a direct action of melatonin on the immune system. Biol Signals 1994;3:107–17.

[445] Potoczek A. Differences in severity and comorbidity of panic and depressive symptoms in difficult and aspirin-induced asthma. Psychiatr Pol 2011;45:469–80.

[446] Pozo D, Reiter RJ, Calvo JR, Guerrero JM. Inhibition of cerebellar nitric oxide synthase and cyclic GMP production by melatonin via complex formation with calmodulin. J Cell Biochem 1997;65:430–42.

[447] Pozo D, Reiter RJ, Calvo JR, Guerrero JM. Physiological concentrations of melatonin inhibit nitric oxide synthase in rat cerebellum. Life Sci 1994;55:455–60.

[448] Priputenova Z. Efficacy of different schemes of aspirin desensitization and the role of arachidonic acid metabolites in the pathogenesis of aspirin-induced bronchial asthma. Synopsis of MD thesis. Moscow: 1990. 25 pp [in Russian].

[449] Priputenova Z, Zebrev A, Chuchalin AG. Eicosanoids and bronchial asthma. Abstracts of papers for the first congress of respiratory diseases. Kiev: 1991. p. 7 [in Russian].

[450] Provinciali M, Di Stefano G, Bulina D, Tibaldi A, Fabris N. Effect of melatonin and pineal grafting on thymocytes apoptosis. Mech Ageing Dev 1996;90:1–19.

[451] Pujols L, Mullol J, Alobid I, Roca-Fererer J, Xaubet A, Picado C. Dynamics of COX-2 in nasal mucosa and nasal polyps from aspirin-tolerant and aspirin-intolerant patients with asthma. J Allergy Clin Immunol 2004;114:814–9.

[452] Pytsky VI, Andrianova NV, Artomasova AV. Allergic diseases. Moscow: Meditsina; 1991. 368 pp [in Russian].

[453] Radomski MW, Palmer RMJ, Moncada S. Glucocorticoids inhibit the expression of an inducible, but not the constitutive nitric oxide synthase in vascular endothelial cells. Proc Natl Acad Sci USA 1990;87:10043–7.

[454] Radomski MW. Biological significance of nitric oxide in platelet function. In: Kubes P, editor. Nitric oxide: a modulator of cell–cell interactions in the microcirculation. R.G. Landes Company; 1995. p. 43–74.

[455] Raevsky KS. Nitric oxide, a new physiological messenger: its possible role in pathology of the central nervous system. Bull Exp Biol Med 1997;123:484–90 [in Russian].

[456] Raihlin NT, Kvetnoy IM, Osadchuk MA. The APUD system: general pathology and oncological aspects. Obninsk 1993;(1):127 [in Russian].

[457] Raihlin NT, Kvetnoy IM, Osadchuk MA. The APUD system: general pathology and oncological aspects. Obninsk 1993;2:108 [in Russian].

[458] Raihlin NT, Kvetnoy IM. Melatonin: extraepiphyseal sources of the hormone in health and disease. Neurobiological aspects of modern endocrinology. Moscow: 1991. p. 49–50 [in Russian].

[459] Randerath WJ. Aspirin-exacerbated respiratory disease. Dtsch Med Wochenschr 2013;138:541–7.

[460] Rawdon BB, Andrew A. Gut endocrine cells in birds: an overview, with particular reference to the chemistry of gut peptides and the distribution, ontogeny, embryonic origin and differentiation of the endocrine cells. Prog Histochem Cytochem 1999;34:3–82.

[461] Redington AE, Meng QH, Springall D, Evans TJ, Créminon C, Maclouf J, et al. Increased expression of inducible nitric oxide synthase and cyclo-oxygenase-2 in the airway epithelium of asthmatic subjects and regulation by corticosteroid treatment. Thorax 2001;56:351–7.

[462] Regodon S, del Prago Miquez M, Jardin I, Lopez JJ, Ramos A, Paredes SD, et al. Melatonin, as an adjuvant-like agent, enhances platelet responsiveness. J Pineal Res 2009;46:275–85.

[463] Regrigny O, Delagrange P, Scalbert E, Lartaud-Idjouadiene I, Atkinson J, Chillon JM. Effects of melatonin on rat pial arteriolar diameter *in vivo*. Br J Pharmacol 1999;127:1666–70.

[464] Reiter RJ. Melatonin biosynthesis, regulation, and effects. In: Stevens RG, Wilson BW, Anderson LE, editors. The melatonin hypothesis. Breast cancer and use of electric power. Richland, WA: Battelle Press; 1997. p. 25–48.

[465] Reiter RJ, Cabrera J, Sainz RM, Mayo JC, Manchester LC, Tan DX. Melatonin as a pharmacological agent against neuronal loss in experimental models of Huntington's disease, Alzheimer's disease and parkinsonism. Ann N Y Acad Sci 1999;890:471–85.

[466] Reiter RJ, Guerrero JM, Garcia JJ, Acuna-Castroviejo D. Reactive oxygen intermediates, molecular damage, and aging. Relation to melatonin. Ann N Y Acad Sci 1998;854:410–24.

[467] Reiter RJ, Maestroni GJM. Melatonin in relation to the antioxidative defense and immune systems: possible implications for cell and organ transplantation. J Mol Med 1999;77:36–9.

[468] Reiter RJ, Tan D-X, Cabrera J, D'Arpa D, Sainz RM, Mayo JC, et al. The oxidant/antioxidant network: role of melatonin. Biol Signals Recept 1999;8:56–63.

[469] Reiter RJ, Tan D-X, Qi W-B. Suppression of oxygen toxicity by melatonin. Acta Pharm Sinica 1998;19:575–81.

[470] Reiter RJ, Tan D-X. Melatonin an antioxidant in edible plants. Ann N Y Acad Sci 2002;957:341–4.

[471] Reiter RJ, Tang L, Garcia JJ, Munoz-Hoyos A. Pharmacological actions of melatonin in oxygen radical pathophysiology. Life Sci 1997;60:2255–71.

[472] Reiter RJ. Melatonin and human reproduction. Ann Med 1998;30:103–8.

[473] Reiter RJ. Oxidative damage in the central nervous system: protection by melatonin. Prog Neurobiol 1998;56:359–84.

[474] Reiter RJ. Pineal function in the human: implications for reproductive physiology. J Obstet Gynaecol 1986;6:77.

[475] Reiter RJ. Pineal melatonin production: photoperiodic and hormonal influences. Adv Pineal Res 1986;1:77–87.

[476] Reiter RJ. Pineal melatonin: cell biology of its synthesis and of its physiological interactions. Endocr Rev 1991;86:151–79.

[477] Reiter RJ, Tan D-X, Kim SJ, Qi W-B. Melatonin as a pharmacological agent against oxidative damage to lipids and DNA. Proc West Pharmacol Soc 1998;41:229–36.

[478] Ressmeyer AR, Mayo JC, Zelosko V, Sainz RM, Tan DX, Poeggeler B, et al. Antioxidant properties of the melatonin metabolite N^1-acetyl-5-methoxykynuramine (AMK): scavenging of free radicals and prevention of protein destruction. Redox Rep 2003;8:205–13.

[479] Reutov VP, Kositsin NS, Svinov MM, Ionkins EG. Problems of neurocybernetics. Rostov-na-Donu: 1995. p. 204–7 [in Russian].

[480] Reutov VP, Sorokina EG, Kositsin NS, Okhotin VE. The problem of nitric oxide in biology and medicine and the principle of cyclicity. Moscow: Nauka; 2003. 96 pp [in Russian].

[481] Reutov VP. Cyclic conversions of nitric oxide in mammals. Moscow: Nauka; 1997. 153 pp [in Russian].

[482] Reutov VP. The nitric oxide cycle in mammals. Achiev Biol Chem 1995;35:189–228 [in Russian].

[483] Ribelayga C, Garidou ML, Malan A, Gauer F, Calgari C, Pévet P, et al. Photoperiodic control of the rat pineal arylakylamine-N-acetyltransferase and hydroxyindole-O-methyltransferase gene expression and its effect on melatonin synthesis. J Biol Rhythms 1999;14:105–15.

[484] Roca-Ferrer J, Perez-Gonzalez M, Garcia-Garcia FJ, Pereda J, Pujols L, Alobid I, et al. Low prostaglandin E (2) and cyclooxygenase expression in nasal mucosa fibroblasts of aspirin-intolerant asthmatics. Respirology 2013;:10.1111/resp.12076.

[485] Romano M, Chen X-S, Takahashi Y, Yamamoto S, Funk CD. Lipoxin synthase activity of human platelet 12-lipoxygenase. Biochem J 1993;296:127–33.

[486] Rom-Bugoslavsksya ES. The role of melatonin in regulation of the endocrine system. Probl Endocrinol 1981;27:81–9 [in Russian].

[487] Rossi MT, Di Bella L, Gualano L, Scalera G. Il sistema megacariociti-plastrine, organo bersaglio della melatonina (MLT). Bull Soc It Biol Sper 1981;17:25–6.

[488] Rossi MT, Di Bella L. Melatonin in thrombocytogenesis. In: Gupta D, Attanasio A, Reiter R, editors. The pineal gland and cancer. Tubingen, London: Brain Research Promotion; 1988. p. 183–94.

[489] Rubin RT, Heist EK, Mcgeoy SS, Hanada K, Lesser IM. Neuroendocrine aspects of primary endogenous depression. XI. Serum melatonin measures in patients and matched control subjects. Arch Gen Psychiatry 2004;49:558–67.

[490] Ryabova KG, Grigoryan GYu. The role of α-adrenergic reception in bronchial asthma. Ther Arch 1988;50:151–6 [in Russian].

[491] Sabry EY. Relation of perimenstrual asthma with disease severity and other allergic—comorbidities—the first report of perimenstrual asthma prevalence in Saudi Arabia. Allergol Immunopathol (Madr) 2011;39:23–6.

[492] Sack RL, Lew J, Hoban TM. Free running melatonin rhythms in blind people: phase shifts with melatonin and Triazolam administration. In: Rensing LV, Mackey MC, editors. Temporal disorder in human oscillatory systems. Heidelberg: Springer-Verlag; 1987. p. 219–24.

[493] Sakai K, Fafeur V, Vulliez-le, Normand B, Dray F. 12-Hydroperoxyeicosatetraenoic acid (12-HPETE) and 15-HPETE stimulate melatonin synthesis in rat pineals. Prostaglandins 1988;35:969–76.

[494] Salvi A, Carrupt P, Tillement J, Testa B. Structural damage to proteins caused by free radicals: assessment, protection by antioxidants, and influence of protein binding (1). Biochem Pharmacol 2000;61:1237–42.

[495] Sampson AP. Leukotriene C4synthase: the engine of aspirin intolerance? Clin Exp Allergy 2011;41:1050–3.

[496] Samter M, Beers RF. Intolerance to aspirin. Clinical studies and consideration of its pathogenesis. Ann Intern Med 1968;68:975–83.

[497] Samuelson B. Leukotrienes: mediators of allergic reactions and inflammation. Int Arch Allergy Appl Immunol 1981;66:96–106.

[498] Sanak M, Kielbasa B, Bochenek G, Szczeklik A. Exhaled eicosanoids following oral aspirin challenge in asthmatic patients. Clin Exp Allergy 2004;34:1899–904.

[499] Sanak M, Levy BD, Clish CB, Chiang N, Gronert K, Mastalerz L, et al. Aspirin-tolerant asthmatics generate more lipoxins than aspirin-intolerant asthmatics. Eur Respir J 2000;16:44–9.

[500] Sanak M, Pierzchalska M, Bazan-Socha S, Szczeklik A. Enhanced expression of the leuko-triene C(4) synthase due to overactive transcription of an allelic variant associated with aspirin-intolerant asthma. Am J Respir Cell Mol Biol 2000;23:273–6.

[501] Sanak M, Sampson AP. Biosynthesis of cysteinyl-leucotrienes in aspirin-intolerant asthma. Clin Exp Allergy 1999;29:306–13.

[502] Sanchez-Borges M, Capriles-Hulett A, Caballero-Fonseca F. The multiple faces of nonste-roidal anti-inflammatory drug hypersensitivity. J Invest Allergol Clin Immunol 2004;14:329–34.

[503] Santos GC, Zucoloto S. Gastrointestinal endocrine cells: brief history and main identifica-tion methods under light microscopy. Arg Gastroenterol 1996;33:36–49.

[504] Satake N, Oe H, Sawada T, Shibata S. The mode of vasorelaxing action of melatonin in rabbit aorta. Gen Pharmacol 1991;22:219–21.

[505] Satouchi M, Maeda H, Yu Y, Yokoyama M. Clinical significance of the increased peak levels of exhaled nitric oxide in patients with bronchial asthma. Intern Med 1996;35:270–5.

[506] Schachter J. Epidemiology of human chlamydial infections. In: Saikki P, editor. Proceedings of the fourth meeting of the European society for *Chlamydia* research. Helsinki: 2000. p. 307–10.

[507] Schaefer D, Gode UC, Baenkler HW. Dynamics of eicosanoids in peripheral blood cells during bronchial provocation in aspirin-intolerant asthmatics. Eur Respir J 1999;13:638–46.

[508] Schedin U, Frostell C, Persson MG, Jakobsson J, Andersson G, Gustafsson LE. Contribution from upper and lower airways to exhaled endogenous nitric oxide in humans. Acta Anaesthesiologica Scandinavica 1995;39:327–32.

[509] Schiappoli M, Gani F, Frati F, Marcucci F, Senna G. Asthma and aspirin. Recenti Prog Med 2003;94:79–87.

[510] Schmidt HW, Hofmann H, Ogilvie P. The role of nitric oxide in physiology and pathology. Heidelberg: Springer; 199575–86

[511] Schmitz M, Szczeklik A. Pathogenesis and pharmacology of aspirin-induced asthma— T-Lymphocytes and eosinophils in asthma pathogenesis. Atemw Lungenkrkh Jahrgang 1994;20:136–41.

[512] Schmitz-Schumann IM, Brauner V, Bode E. The IgE and IgG antibody spectrum against non-steroidal anti-inflammatory drugs in aspirin induced asthma. Eur Respir J 1989;2:274.

[513] Schuster C. Sites and mechanisms of action of melatonin in mammals: the MT1 and MT2 receptors. J Soc Biol 2007;201:85–96.

[514] Schwartz HJ, Bennett B. The differential effects of acetylsalicylic acid on *in vitro* aggrega-tion of platelets from normal, asthmatic and aspirin-sensitive subjects. Int Arch Allergy Appl Immunol 1973;45:899–904.

[515] Seabra B, Duarte R, Sa RC. Asthma, nasal polyposis and aspirin intolerance—a triad to remember. Rev Port Pneumol 2006;12:709–14.

[516] Seidov VD, Alekberadze AV. The importance of morphofunctional state of apudocytes for predicting hemorrhage in duodenal ulcer. Surgery 2000;(9):16–9 [in Russian].

[517] Sener A, Ozsavci C, Bingol-Ozakpinar O, Cevik O, Yanikkaya-Demirel G, Yardimci T. Oxidized-LDL and Fe^{3+}/ascorbic acid-induced oxidative modifications and phosphatidyl-serine exposure in human platelets are reduced by melatonin. Folia Biol (Praha) 2009;55:45–52.

[518] Serebrykiva VI, Shabrov AV, Litvinov AS. Clinical and pathogenic peculiarities of neuro-endocrine regulation in case bronchial obstruction and arterial hypertension. Abstracts of

the seventh national congress of respiratory diseases. Moscow: 1997; No. 1610. p. 432 [in Russian].

[519] Serhan CH. Lipoxins and aspirin-triggered 15-epi-lipoxins are the first lipid mediators of endogenous antiinflammation and resolution. Prostaglandins Leukot Essent Fatty Acids 2005;73:141−62.

[520] Serhan CN, Chiang N. Lipid-derived mediators in endogenous anti-inflammation and resolution: lipoxins and aspirin-triggered 15-epi-lipoxins. Sci World J 2002;2:169−204.

[521] Serhan CN, Sheppard K-A. Lipoxin formation during human neutrophil−platelet interactions (evidence for the transformation of leukotriene A_4 by platelet 12-lipoxygenase in vitro). J Clin Invest 1990;85:772−80.

[522] Sessa WC. The nitric oxide synthase family of proteins. J Vasc Res 1994;31:131−43.

[523] Shaikh AY, Xu J, Wu Y. Melatonin protects bovine cerebral endothelial cells from hyperoxia-induced DNA damage and death. Neurosci Lett 1997;229:193−7.

[524] Shi J, Misso NL, Duffy DL, Bradley B, Beard R, Thompson PJ, et al. Cyclooxygenase-1 gene polymorphisms in patients with different asthma phenotypes and atopy. Eur Respir J 2005;26:249−56.

[525] Shibata S, Satake N, Takagi T, Usui H. Vasorelaxing action of melatonin in rabbit basilar artery. Gen Pharmacol 1989;20:677−80.

[526] Shmushkovich BI, Cheglakova TA, Chuchalin AG. Bronchial asthma: mechanisms of corticoid dependence. Pulmonology 1993;(1):35−48 [in Russian].

[527] Shrestha Palikhe N, Kim SH, Jin HJ, Hwang EK, Nam YH, Park HS. Genetic mechanisms in aspirin-exacerbated respiratory disease. J Allergy (Cairo) 2012;2012:794890. Available from: http://dx.doi.org/10.1155/2012/794890 Epub 2011 Aug 7.

[528] Shustov SB, Havinson VH, Shutak TS, Romashevsky BV. The action of epithalamin on the carbohydrate metabolism and condition of the cardiovascular system in patients with insulin-dependent diabetes. Clin Med 1998;76:45−8 [in Russian].

[529] Silva CLM, Tamura EK, Macedo SMD, Cecon E, Bueno-Alves L, Farsky SHP, et al. Melatonin inhibits nitric oxide production by microvascular endothelial cells in vivo and in vitro. Br J Pharmacol 2007;151:195−205.

[530] Simko M, Mattsson MO. Extremely low frequency electromagnetic fields as effectors of cellular responses in vitro: possible immune cell activation. J Cell Biochem 2004;93:83−92.

[531] Simon MI, Strathmann MP, Gautam N. Diversity of G proteins in signal transduction. Science 1991;252:802−8.

[532] Sjoblom M, Jedstedt G, Flemstrom G. Peripheral melatonin mediates neural stimulation of duodenal mucosal bicarbonate secretion. J Clin Invest 2001;108:625−33.

[533] Sladek K, Dworski R, Soja J, Sheller JR, Nizankowska E, Oates JA, et al. Eicosanoids in bronchoalveolar lavage fluid of aspirin-intolerant patients with asthma after aspirin challenge. Am J Respir Crit Care Med 1994;149:940−6.

[534] Smirnov AN. Nuclear melatonin receptors. Biochemistry 2000;66:19−26.

[535] Smith CM, Hawksworth RJ, Thien FC, Christie PE, Lee TH. Urinary leukotriene E_4 in bronchial asthma. Eur Respir J 1992;5:693−9.

[536] Sneddon JM, Vane JR. Endothelium-derived relaxing factor reduces platelet adhesion to bovine endothelial cells. Proc Natl Acad Sci U S A 1988;85:2800−4.

[537] Sodin-Semri S, Taddeo B, Tseng D, Varga J, Fiore S. Lipoxin A_4 inhibits IL-1β-induced IL-6, IL-8, and matrix metalloproteinase-3 production in human synovial fibroblasts and enhances synthesis of tissue inhibitors of metalloproteinases. J Immunol 2000;164:2660−6.

[538] Sokovnina YaM. Megakaryocytes: metabolism and function. Achiev Mod Biol 1992;112:541–53 [in Russian].

[539] Sousa A, Parikh A, Scadding G, Corrigan CJ, Lee TH. Leukotriene-receptor expression on nasal mucosal inflammatory cells in aspirin-sensitive rhinosinusitis. N Engl J Med 2002;347:1524–6.

[540] Sousa AR, Lams BE, Pfister R, Christie PE, Schmitz M, Lee TH. Expression of interleukin-5 and granulocyte-macrophage colony-stimulating factor in aspirin-sensitive and non-aspirin-sensitive asthmatic airways. Am J Respir Crit Care Med 1997;156:1384–9.

[541] Sousa AR, Pfister R, Christie PE, Lane SJ, Nasser SM, Schmitz-Schumann M, et al. Enhanced expression of cyclo-oxygenase isoenzyme 2 (COX-2) in asthmatic airways and its cellular distribution in aspirin-sensitive asthma. Thorax 1997;52:940–5.

[542] Stahl SM. Platelets as pharmacologic models for the reception and biochemistry of mono-aminergic neurons. In: Longenecker GL, editor. The platelets physiology and pharmacology. New York, NY: 1985. p. 308–39.

[543] Stehle JH. Pineal gene expression: dawn in a dark matter. J Pineal Res 1995;18:179–90.

[544] Steinhilberg D, Brungs M, Werz O, Wiesenberg I, Danielsson C, Kahlen JP, et al. The nuclear receptor for melatonin represses 5-lipoxygenase gene expression in human B lymphocytes. J Biol Chem 1995;270:7037–40.

[545] Stevenson DD, Sanchez-Borges M, Szczeklk A. Classification of allergic and pseudoallergic reaction to drugs that inhibit cyclooxygenase enzymes. Ann Allergy Asthma Immunol 2001;87:177–80.

[546] Stevenson DD, Szczeklik A. Clinical and pathologic perspectives on aspirin sensitivity and asthma. J Allergy Clin Immunol 2006;118:773–85.

[547] Stirling RG, Van Rensen ELJ, Barnes PJ, Chung KF. Interleukin-5 induces CD34+ eosinophil progenitor mobilization and eosinophil CCR3 expression in asthma. Am J Respir Crit Care Med 2001;164:1403–9.

[548] Strada SJ, Weiss B. Increase response to catecholamines of the cyclic AMP system of rat pineal gland induced by decreased sympathetic activity. Arch Biochem Biophys 1974;160:97.

[549] Sumarokov AB, Lyakishev A. Chlamydial infection caused by C. pneumonia and atherosclerosis. Clin Med 1998;76:4–10 [in Russian].

[550] Sumarokov AB, Pankratova VN, Lyakisheva AA, Avdeeva IYu. Studying Chlamydia pneumoniae in atherosclerosis. Clin Med 1999;77:4–7.

[551] Suzdaltseva TV. An integrated approach to the evaluation of immune reactivity of bronchial asthmatics. Synopsis of MD thesis. Leningrad: 1988. 18 pp [in Russian].

[552] Suzdaltseva TV. Immunopathological aspects of aspirin-induced bronchial asthma. Allergology 1999;(4):16–8 [in Russian].

[553] Suzuki K, Hasegawa T, Sakagami T, Koya T, Toyabe S, et al. Analysis of perimenstrual asthma based on questionnaire surveys in Japan. Allergol Int 2007;56:249–55.

[554] Swerkosz TA, Mitchell JA, Warner TD, Botting RM, Vane JR. Co-induction of nitric-oxide synthase and cyclo-oxygenase: interactions between nitric oxide and prostanoids. Br J Pharmacol 1995;114:1335–42.

[555] Syromyatnikova NV, Goncharova VA, Kotenko TV. The metabolic activity of the lungs. Leningrad: Meditsina; 1987. 167 pp [in Russian].

[556] Szab C, Ohshima H. DNA damage induced by peroxynitrite: subsequent biological effects. Nitric Oxide 1997;1:373–85.

[557] Szczeklik A, Gryglewski RJ, Czerniawska-mysik G. Relationship of inhibition of prosta-glandin biosynthesis by analgesics to asthma attacks in aspirin-sensitive patients. Br Med J 1975;1:67−9.

[558] Szczeklik A, Nisankowska-Mozankowska E, Sanak M. Hypersensitivity to aspirin and nonsteroidal anti-inflammatory drugs. In: Adkinson NF, Busse WW, Bochner BS, Holgate ST, Simons FGR, Lemanske RF, editors. Middleton's allergy. 7th ed. Philadelphia, PA: Mosby, Elsevier; 2009. p. 1227−43.

[559] Szczeklik A, Nizankowska E, Bochenek G, Nagraba K, Mejza F, Swierczynska M. Safety of a specific COX-2 inhibitor in aspirin-induced asthma. Clin Exp Allergy 2000;31:219−25.

[560] Szczeklik A, Nizankowska E, Matalerz L, Szabo Z. Analgesics and asthma. Am J Ther 2002;9:233−43.

[561] Szczeklik A, Nizankowska E. Clinical features and diagnosis of aspirin induced asthma. Thorax 2000;55:42−4.

[562] Szczeklik A, Picado C. Aspirin-induced asthma. Chung F, Fabbri LM, editors. Europ. Respir. Monograph, 2003;8:239−59.

[563] Szczeklik A, Sanak M. The broken balance in aspirin hypersensitivity. Eur J Pharmacol 2006;533:145−55.

[564] Szczeklik A, Sanak M. The role of COX-1 and COX-2 in asthma pathogenesis and its sig-nificance in the use of selective inhibitors. Clin Exp Allergy 2002;32:339−42.

[565] Szczeklik A, Schmitz-Schumann M. Aspirin-induced asthma: from pathogenesis to therapy. Allergol Immunopathol 1993;21:35−40.

[566] Szczeklik A, Sladek K, Dworski R, Nizankowska E, Soja J, Oates J. Bronchial aspirin challenge causes specific eicosanoid response in aspirin-sensitive asthmatics. Am J Respir Crit Care Med 1996;154:1608−14.

[567] Szczeklik A, Stevenson DD. Aspirin-induced asthma: advances in pathogenesis and man-agement. J Allergy Clin Immunol 1999;104:5−13.

[568] Szczeklik A. About aspirin-induced asthma. Newsletter 1994;2.

[569] Szczeklik A. Aspirin-induced asthma: a tribute to John Vane as a source of inspiration. Pharmacol Rep 2010;62:526−9.

[570] Szczeklik A. Aspirin-induced asthma: pathogenesis and clinical presentation. Allergy Proc 1992;13:163−73.

[571] Szczeklik A. Aspirin-tolerant asthmatics generate more lipoxins than aspirin-intolerant asthmatics. Eur Respir J 2000;16:44−9.

[572] Szczeklik A. Mechanisms of aspirin-induced asthma. In: Szczeklik A, Gryglewski RJ, Vane JR, editors. Eicosanoids, aspirin, and asthma. New York, NY: 1999. p. 299−315.

[573] Szczeklik A. The cyclooxygenase theory of aspirin-induced asthma. Eur Respir J 1990;3:588−93.

[574] Szulakowski P, Pierzchala W. Clinical aspects of *Chlamydia* respiratory tract infections and their role in the pathogenesis of asthma. Wiad Lek 1998;51:202−7.

[575] Takano T, Fiore S, Maddox JF, Brady HR, Petasis NA, Serhan CN. Aspirin-triggered 15-epi-lipoxin A_4(LXA$_4$) and LXA$_4$ stable analogues are potent inhibitors of acute inflamma-tion: evidence for anti-inflammatory receptors. J Exp Med 1997;85:1693−704.

[576] Tan DX, Manchester LC, Burkhardt S, Sainz RM, Mayo JC, Kohen R, et al. N^1-acetyl-N^2-formyl-5-methoxykynuramine, a biogenic amine and melatonin metabolite, functions as a potent antioxidant. FASEB J 2001;15:2294−6.

[577] Tan D-X, Manchester LC, Reiter RJ, Plummer BF. Cyclic 3-hydroxymelatonin: a melatonin metabolite generated as a result of hydroxyl radical scavenging. Biol Signals Recept 1999;8:70−4.

[578] Tan D-X, Manchester LC, Reiter RJ, Qi WB, Zhang M, Weintraub ST, et al. Identification of highly elevated levels of melatonin in bone marrow its origin and significance. Biochim Biophys Acta 1999;1472:206−14.

[579] Tatarsky AR, Emirova S. The role of platelets in the pathogenesis of some forms of bronchial asthma. Ther Arch 1990;62:149−51 [in Russian].

[580] Longenecker GL, editor. The platelets. Physiology and pharmacology. Toronto, Montreal, Sydney, Tokyo: Academic Press; 1985. p. 489.

[581] Thien FCK, Walters EH. Eicosanoids and asthma: an update. Prostaglandins Leukot Essent Fatty Acids 1995;52:271−88.

[582] Tolga UZ, Manev H. Circadian expression of pineal 5-lipoxygenase mRNA. Neuroreport 1998;9:783−6.

[583] Torres-Farfan C, Valenzuela FJ, Mondaca M, Valenzuela GJ, Krause B, Herrera EA, et al. Evidence of a role for melatonin in fetal sheep cerebral artery, brown adipose tissue and adrenal gland. J Physiol 2008;586:4017−27.

[584] Touitou I, Haus E. Circannual variation of some endocrine and neuroendocrine functions in humans. In: Wetterberg L, editor. Light and biological rhythms in man. Stockholm: 1993. pp. 313−27.

[585] Touitou Y, Bogdan A, Haus E, Touitou C. Modifications of circadian and circannual rhythms with aging. Exp Gerontol 1997;32:603−14.

[586] Trinus FP, Mohort NA, Klebanov BM. Non-steroid anti-inflammatory drugs. Kiev: Zdorovya; 1975. 240 pp [in Russian].

[587] Trofimov VI, Evsyukova EV, Bondarenko VL, Katysheva NS. Glucocorticoid function of the adrenal gland and the level of melatonin in patients with aspirin-induced asthma. Pulmonology 1998;(2):68−70 [in Russian].

[588] Trofimov VI, Shaporova NL, Dudina OV. Hormone-dependent bronchial asthma: pathogenesis, clinical findings and treatment. IPP-SPSMU Record 2001;8:52−6 [in Russian].

[589] Trofimov VI, Shaporova NL, Lebedeva DP. Hormonal activity disorders in the adrenal cortex and ovary of bronchial asthmatics. Ther Arch 1991;63:75−9 [in Russian].

[590] Trofimov VI, Vishnevskaya NL. On some hormonal regulation disorders in bronchial asthmatics. Ther Arch 1989;61:89−91 [in Russian].

[591] Trofimov VI. The role of dyshormonal disorders in the formation of inflammation. Mechanisms of inflammation of the bronchi and lungs and anti-inflammatory therapy. Ed. by GB. Fedoseev, 1998. St. Petersburg: Nordmedizdat; 1998. p. 387−409 [in Russian].

[592] Tslm ST, Wong JT, Wong YH. CGP 52608-induced cyst formation in dinoflagellates: possible involvement of a nuclear receptor for melatonin. J Pineal Res 1996;21:101−7.

[593] Tuev AV, Mishlanov VYu. Bronchial asthma: immunity, hemostasis, treatment. Perm: IPK Zvezda; 2000. 220 pp [in Russian].

[594] Uchida K, Aoki T, Satoh H, Tajiri O. Effects of melatonin on muscle contractility and neuromuscular blockade produced by muscle relaxants. Jpn J Anesthesiol 1997;46:205−12.

[595] Umbreit C, Virchow JC, Thorn C, Hormann K, Klimek L, Pflaar O. Aspirin-intolerance-syndrome: a common and interdisciplinary disease. Internist (Berl) 2010;51:1196−8.

[596] Uneri C, Ozturk O, Polat S, Yuksel M, Haklar G. Determination of reactive oxygen in nasal polyps. Rhinology 2005;43:185−9.

[597] Urazaev AH, Zefirov AL. The physiological role of nitric dioxide. Achiev Physiol 1999;30:54−71 [in Russian].

[598] Usoltseva NV, Usoltsev VA. Liquid crystal state of biologic structures as a necessary factor of metabolism. Liquid Crystals Ivanovo 1978;11−34 [in Russian].

[599] Uz T, Longone P, Manev H. Increased hippocampal 5-lipoxygenase mRNA content in melatonin-deficient, pinealectomized rats. J Neurochem 1997;69:2220−3.

[600] Vacas MI, Del Zar MM, Martinuzzo M, Cardinali DP. Binding sites for (3H)-melatonin in human platelets. J Pineal Res 1992;13:60−5.

[601] Vacas MI, Del Zar MM, Martinuzzo M, et al. Inhibition of human platelet aggregation and thromboxane B2 production by melatonin correlation with plasma melatonin levels. J Pineal Res 1991;11:135−9.

[602] Valevski A, Modai I, Jerushalmy Z, Kikinzon L, Weizman A. Effect of melatonin on active transport of serotonin into blood platelets. Psychiatry Res 1995;57:193−6.

[603] Vane JR. Inhibition of prostaglandin synthesis as a mechanism of action for aspirin-like drugs. Nature 1971;231:232−4.

[604] Vanecek J, Vollrath L. Localization and characterization of melatonin receptors. In: Reiter RJ, Lukaszyk A, editors. Advances in pineal research, 4. London: John Libbey Company; 1990. p. 147−54.

[605] Vanecek J. Melatonin inhibits increase of intracellular calcium and cyclic AMP in neonatal rat pituitary via independent pathways. Mol Cell Endocrinol 1995;107:149−53.

[606] Vasan R, Beder I, Styk J. Melatonin and the heart. Cesk Fysiol 2004;53:29−33.

[607] Violi F, Marino R, Milite MT, Loffredo L. Nitric oxide and its role in lipid peroxidation. Diabetes Metab Res Rev 1999;15:283−7.

[608] Vishnyakova LA. The role of various microorganisms and infectious processes in the occurrence and course of bronchial asthma. Ther Arch 1990;62:57−62 [in Russian].

[609] Vlasov TD, Vivulanets EV, Mindukshev IV. Functional activity of platelets in ischemia/reperfusion of the rat brain. Sechenov Russ Physiol J 2000;86:422−6 [in Russian].

[610] Von Gall C, Stehle JH, Weaver DR. Mammalian melatonin receptors: molecular biology and signal transduction. Cell Tissue Res 2002;309:151−62.

[611] Wakatsuki A, Okatani Y, Izumiya C, Ikenoue N. Melatonin protects against ischemia and reperfusion-induced oxidative lipid and DNA damage in fetal rat brain. J Pineal Res 1999;26:147−52.

[612] Wan KS, Wu WF. Eicosanoids in asthma. Acta Pediatr Taiwan 2007;48:299−304.

[613] Wang M, Yokotani K, Nakamura K, Murakami Y, Okada S, Osumi Y. Melatonin inhibits the central sympatho-adrenomedullary outflow in rats. Jpn J Pharmacol 1999;81:29−33.

[614] Wang WZ, Fang XH, Sephenson LL, Baynosa RC, Zamboni WA. Microcirculatory effects of melatonin in rat skeletal muscle after prolonged ischemia. J Pineal Res 2005;39:57−65.

[615] Wang XS, Wu AY, Leung PS, Lau HY. PGE suppresses excessive anti−IgE induced cysteinyl leucotrienes production in mast cells of patients with aspirin exacerbated respiratory disease. Allergy 2007;62:6320−627.

[616] Wang YT, Chen SL, Xu SY. Effect of melatonin on the expression of nuclear factor-kappa B and airway inflammation in asthmatic rats. Zhonghua Er Ke Za Zhi 2004;42:94−7.

[617] Wanke IE. Reproduction and the APUD system. Semin Surg Oncol 1993;9:394−8.

[618] Ward JK, Belvisi MG, Fox AJ, Miura M, Tadjkarimi S, Yacoub MH, et al. Modulation of cholinergic neural bronchoconstriction by endogenous nitric oxide and vasoactive intestinal peptide in human airways *in vitro*. J Clin Invest 1993;92:736−42.

[619] Wardlay AJ, Hay H, Cromwell O, Collins JV, Kay AB. Leukotrienes, LTC4 and LTB4 in bronchoalveolar lavage in bronchial asthma and other respiratory disease. J Allergy Clin Immunol 1989;84:19−26.

[620] Webb DJ, Haynes WG. Venoconstriction to endothelin-1 in humans is attenuated by local generation of prostacyclin but not nitric oxide. J Cardiovasc Pharmacol 1993;22:17−20.

[621] Week A, Sanz ML, Gamboa P. New pathophysiological concepts on aspirin hyper sensitivity (Widal syndrome); diagnostic and therapeutic consequences. Bull Acad Natl Med 2005;189:1201−18.

[622] Weekley LB. Effect of melatonin on pulmonary and coronary vessels are exerted through perivascular nerves. Clin Auton Res 1993;3:45−7.

[623] Weekley LB. Influence of melatonin on bovine pulmonary vascular and bronchial airway smooth muscle tone. Clin Auton Res 1995;5:53−6.

[624] Weekly LB. Effects of melatonin on isolated pulmonary artery and vein: role of the vascular endothelium. Pulm Pharmacol 1993;6:149−54.

[625] Wei X-Q, Charles IG, Smith A, Ure J, Feng GJ, Huang FP, et al. Altered immune responses in mice lacking inducible nitric oxide synthase. Nature 1995;375:408−11.

[626] Wetterberg L. Light and melatonin in humans. In: Stevens RG, Wilson BW, Anderson LE, editors. The melatonin hypothesis. Breast cancer and use of electric power. Richland, WA: Battelle Press; 1997. p. 233−65.

[627] White AA, Stevenson DD. Aspirin-exacerbated respiratory disease: update on pathogenesis and desensitization. Semin Respir Crit Care Med 2012;33:588−94.

[628] White DG, Mundin JW, Summer MJ, Watts IS. The effect of endothelins on nitric oxide and prostacyclin production from human umbilical vein, porcine aorta and bovine carotid artery endothelial cells in culture. Br J Pharmacol 1999;109:1128−32.

[629] Williams FM, Asad SI, Lessof MH, Rawlins MD. Plasma esterase activity in patients with aspirin-sensitive asthma or urticaria. Eur J Clin Pharmacol 1987;33:387−90.

[630] Wolfe RN, Hui KK, Conolly ME, Tashkin DP, Fisher HK. A study of beta-adrenergic and prostaglandin receptors in patients with aspirin-induced bronchospasm. J Allergy Clin Immunol 1982;69:46−50.

[631] Wttrly LB. Pharmacologic studies on the mechanism of melatonin induced vasorelaxation in rat aorta. J Pineal Res 1995;19:133−8.

[632] Wu CS, Leu SF, Yang HY, Huang BM. Melatonin inhibits the expression of steroidogenic acute regulatory protein and steroidogenesis in MA-10 cells. J Androl 2001;22:245−54.

[633] Yakovlev VA, Trofimov VM, Vavilov AG. Selected problems of clinical endocrinology. St. Petersburg: Orgtekhizdat; 1995. 130 pp [in Russian].

[634] Yakovleva NV. Causes of inflammation. Respiratory viruses. In: Fedoseev Prof. GB, editor. Mechanisms of inflammation of the lungs and bronchi and anti-inflammatory therapy. St. Petersburg: Nordmedizdat; 1998. p. 25−66 [in Russian].

[635] Yalkut CI, Petrovskaya IA, Evseeva TA. Investigation into the peculiarities of asthma triad pathogenesis and clinic. Ther Arch 1979;21:46−9 [in Russian].

[636] Yamaguchi H, Higashi N, Mita H, Ono E, Komase Y, Nakagawa T, et al. Urinary concentrations of 15-epimer of lipoxin A(4) are lower in patients with aspirin-intolerant compared with aspirin-tolerant asthma. Clin Exp Allergy 2011;41:1711−8.

[637] Yamamoto S. Mammalian lipoxygenases: molecular and catalytic properties. Prostaglandins Leukot Essent Fatty Acids 1989;35:219−29.

[638] Yamashita T, Tsuyi H, Maeda N, Tomoda K, Kumazawa T. Etiology of nasal polyps associated with aspirin-sensitive asthma. Rhinology 1989;8:15−24.

[639] Yan MY, Pang CS, Kravtsov G, Pang SF, Shiu SY. 2[125]iodomelatonin sites in guinea pig platelets. J Pineal Res 2002;32:97–105.

[640] Yildiz M, Akdemur O. Assessment of the effects of physiological release of melatonin on arterial distensibility blood pressure. Cardiol Young 2009;19:198–203.

[641] Ying S, Meng Q, Scadding G, Parikh A, Corrigan CJ, Lee TH. Aspirin-sensitive rhinosinusitis is associated with reduced E-prostanoid 2 receptor expression. J Allergy Clin Immunol 2006;117:312–8.

[642] Yoshimura T, Yoshikawa M, Otori N, Haruna S, Moriyama H. Correlation between the prostaglandin D2/E2 ratio in nasal polyps and the recalcitrant pathophysiology of chronic rhinosinusitis associated with bronchial asthma. Allergol Int 2008;57:429–36.

[643] Yu C, Zhu XuS. The analgesic effects of peripheral and central administration of melatonin in rats. Eur J Pharmacol 2000;403:49–53.

[644] Zamorskyi II, Pishak VP, Meshchyshen IF. The effect of melatonin on photoperiod changes in the glutathione system of the brain under acute hypoxia. Fiziol Zh 1999;45:69–76.

[645] Zapolska-Downar B. Silver-absorptive cells in liver parenchyma.II. Morphochemical studies in guinea pigs after anaphylactic and histamine-induced shock. J Cardiovasc Pharmacol 1985;7:79–85.

[646] Zarudy FS. Arachidonates as bronchial tonus regulators. Pathophysiol Exp Ther 1989; (2):71–9 [in Russian].

[647] Zemskov AM, Zemskov VM, Zoloedov VI, Bzhozovsky E. Immune reactivity as a factor of body homeostasis regulation. Achiev Mod Biol 1999;119:99–114 [in Russian].

[648] Zhang H, Akbar M, Kim HY. Melatonin: an endogenous negative modulator of 12-lipoxygenation in the rat pineal gland. Biochem J 1999;344:487–93.

[649] Zhang QZ, Zhang JT. Inhibitory effects of melatonin on free intracellular calcium in mouse brain cells. Chung Kuo Yao Li Hsueh Pao 1999;20:206–10.

[650] Zhu D, Zhao J, Mo L, Li H. Drug allergy: identification and characterization of IgE-reactivities to aspirin and related compounds. J Invest Allergol Clin Immunol 1997;7:160–8.

[651] Zietkowski Z, Bodzenta-Lukaszyk A, Tomasiak MM, Skiepko R, Szmitkowski M. Comparison of exhaled nitric oxide measurement with conventional tests in steroid-naive asthma patients. J Investig Allergol Clin Immunol 2006;16:239–46.

[652] Zonis BY. The role and the adrenoreceptor apparatus in bronchospasm mechanisms and diagnostic methods for different type of adrenergic imbalance in bronchial asthmatics. Ther Arch 1989;61:43–6 [in Russian].

www.ingramcontent.com/pod-product-compliance
Lightning Source LLC
Chambersburg PA
CBHW070731220326
41598CB00024BA/3385